QUEST FOR THE

INNER EARTH

by □ orothy Leon

TGS Publishers
2004

Quest for the Inner Earth
Author: Dorothy Leon

Non-fiction, science, suppressed science, history

2004 TGS Publishers
ISBN 0-9742018-5-5

Printed by

TGS PUBLISHERS
22241 Pinedale Lane
Frankston, Texas 75763
903-876-3256
www.HiddenMysteries.com

Other books by Dorothy Leon:

Great Cosmic Council (a channeled epic)
Echoes, Traces and Reflections (poetry)
Triangle From Mountains (ley lines and vortexes)
The Three R's: (Biblical scriptures)
    Reality of the Light
    Reincarnation Reviewed
    Revelation Revealed
Saint Germain Trilogy:
1. Saint Germain's Mystical Quest for the Golden Age
2. Saint Germain's Mystical Quest for Freedom and Democracy
3. Saint Germain and Portia's Mystical Role as Directors of the Aquarian Age

A Mystical Walk Through Nature
Is Jehovah an E.T.? (UFOs in the Bible)
Beyond the Follies of Fundamentalism

DEDICATION

This book is dedicated to truth seekers everywhere. May you find the Inner Earth Paradise that resides within you.

ACKNOWLEDGEMENTS

I would like to express my heart-felt gratitude to Sandra Nelson, Lorae Ireland, and Alice Hestad--my trinity of angels--without whose help, this book could not have been written. I also thank my wonderful friend and talented artist, Mary Scott Christy, and all the daring trailblazers--especially Jan Lamprecht, Michael Mott, Ralph Eppich, Rodney Cluff, David Pratt, Dean De Lucia, Sadek Adam, Maurice Doreal, Dr. Raymond Bernard, and Bruce Walton--who have paved the way for our understanding of this vast subject. I thank Jan Lamprecht for his help in editing the scientific portion of this book and for writing my introduction. I also thank you who are reading this for being open-minded enough to explore this intriguing subject. May all be blessed.

INTRODUCTION

by Jan Lamprecht, author of *Hollow Planets*

Children and young people these days often say, in their ignorance and arrogance, that there is no point in studying history. Only someone who had never actually studied history would dare say something so ignorant.

Knowledge of history's is essential if one is to understand the world today. Indeed, virtually all the greatest leaders of the Western world were students of history. Nobody studies history more avidly than military leaders--past and present--because many of the same tricks, which worked for Alexander the Great, also work today.

Winston Churchill and Napoleon, both great leaders in their time, were avid readers of history. They knew much of the history of Europe going back to ancient times. Both of them questioned many accepted viewpoints. Both of them expressed the view that many things people believed, were in fact not so.

For example, Napoleon once said: "History is lies agreed upon". Winston Churchill said something to the effect that at some point in a man's life, he may stumble upon the truth. He will then pick himself up and pretend it never happened.

Both Churchill and Napoleon were referring to a lot of the "small detail" of history. They were referring to various "accepted facts" which everyone agreed were true, but which in fact were not. They did however agree, that the broad facts, the major trends of history were well worth studying and that many deeper lessons and insights could be gained from this.

Both Churchill and Napoleon had a quirky and open-minded side to them. Churchill, in his earlier life believed in the illuminati (he later changed his mind). Napoleon loved ghost stories and enjoyed telling them to children. Napoleon once commented that without a doubt, Jesus could not have been a normal man.

My own particular life in Southern Africa has caused me to question everything around me and to take nothing at face value. I grew up in Rhodesia (now Zimbabwe) during the most violent period of its history, and I left it, only to end up in South Africa during the most violent period of its history. I was unfortunate (or fortunate?) enough, to live in two countries, which happened to be in the news, around the world, daily. In the 1970's, it was "The Rhodesian Question", and in the 1980's it was "South African Apartheid".

It was strange living in countries, which were the focus of world attention, seeing things happening around me, and then hearing what CNN or the BBC had to say about it. I would hear about the experiences of friends and family, and then see a completely distorted picture of this being presented to the rest of the world by the Western Mass Media.

Perhaps I was unlucky. But I tend to think I was actually blessed. I was blessed in a way that most people in the Western world are not. I knew, from an early age, that a lot of accepted "facts" in this world are not correct. They are easily twisted out of context. I knew, from my life's experiences, that both Churchill and Napoleon were right-- the "detail of history" can be incorrect. The victors write history and they put the slant on it that makes them look like the good guys. Such is life.

I have lived in South Africa for over 20 years, and I can tell you that there is no country on this Earth about which more lies have been told. Like with Zimbabwe, the lies still dominate. I hope that one day the facts about this country too, will come out. For those who are interested, I did write on my experiences in a book called *Government by*

Deception, which is available on the Internet. In it I wrote about propaganda and how it works in the real world.

My life experiences taught me to take nothing at face value. They taught me to question every fact I stumble upon. They taught me not to be intimidated just because a majority of people tell you that you are wrong. Just because everyone else believes it doesn't make it right. My life experiences also taught me to read and read, and to especially read history. Much can be picked from history. And with maturity, one also becomes more adept at spotting those "lies agreed upon" and at drawing new conclusions from history.

Science is no different to other human endeavors. Just because a group of clever men agree upon something does not always make them right. The history of science is replete with ideas, which were once regarded as "fact", and which were later disproved.

I believe one should have an open mind. And one should read many things, but one should be selective about what you believe. The quest for knowledge is never in vain. This is how you grow.

In this book you will find many references to history. Don't expect them all to be true. But be open-minded, and investigate and think for yourself. Be prepared to think the unthinkable and to open your mind to new possibilities. Be prepared to question things. Also, use a book such as this, as a study guide, so that you may study more, and learn more. At all times, think for yourself, and reach your own conclusions.

Science these days is also politicized. Many scientists are atheists. This has colored their outlook on spiritual matters and on the possibility that Religion and Science may converge at some point. Scientists, like all humans, are prone to their own errors. Just because a scientist says something, doesn't mean it is fact.

Churchill experienced this first hand. Intelligence reports in World War II said that the Germans were building what appeared to be rockets. His science advisor told him that the rockets described in the intelligence reports were too small to be real. His scientific advisor was convinced the only workable rockets were solid-fuel ones. But, solid-fuel rockets, which could fly to distance to England, would have to be huge. They would also need big platforms to stand on. So the scientists argued, for years, against this incoming intelligence, continuously debunking it. It was only when the British finally managed to get photographs of the rockets and retrieved parts of a crashed rocket that the scientist eventually admitted he was wrong. Due to his marrow mindedness, he had prevented the British from taking action against the rockets until it was almost too late. What the scientist had overlooked was the possibility that the Germans would be ingenious enough to make a liquid-fuel rocket work. He had not regarded this as being technically possible. But he was wrong.

If scientists can be wrong about the Germans, then how much more likely is it that they are going to be wrong about God"

What lies deep under the earth? We don't really know. The deepest our instruments have gone down is 20 miles on a planet 7,926 miles in diameter. Maybe the interior of our planet, which we think is dead, harbors amazing things. Let us not blind ourselves to these possibilities. Whatever you do, do not believe everything you see and hear in the popular press. The truth could be very different.

When I wrote Hollow Planets (see: www.HollowPlanets,com), I exercised some skepticism towards science. I did my own thinking. I let my mind roam free, and I deliberately sought out the facts, which do not fit. I assembled them in ways most people would never even dare to think of. Though it was published in 1999, I have never yet had any scientist anywhere in the world, write to me and say: "Mr. Lamprecht, you lied about this or that. . you misrepresented these scientific facts". Indeed, I have had a handful of scientists read it and write to me, congratulating me on my book. The majority won't look at it. But nobody, who has actually read it, has ever been able to said I lied, though I bet many would have loved to do so, if they could have found the opportunity.

One of the most bizarre concepts in the Hollow Earth theory is the central sun. It appeared to me, the solution to this problem might be a small-scale nuclear fission reaction. I was the first "popular" writer to ever talk about the possibility that there might be a nuclear reaction at the center of the Earth. Few knew that nuclear scientists had spoken about underground nuclear reactions since the 1940's. Thanks to my book, a nuclear scientist in the USA, Dr. Marvin Herndon, went public and spoke about how science is manipulated, and ideas are suppressed. He wrote about his own studies, which showed that a nuclear reaction 6 miles in diameter must exist at the core of the Earth. In probability, it exists inside most planets. His papers have been published in a number of prominent scientific journals.

In recent years, a number of major magazines have started publishing Dr. Herndon's ideas (see his website: www.NuclearPlanet.com). They refer to it as the "geo-reactor". So the next time you see that term in a newspaper or magazine, remember it was first discussed in *Hollow Planets*. Currently a number of scientists are starting to study Herndon's ideas. Most of the research is into neutrinos, which are nuclear particles, which, amazingly, are rising up from inside the Earth. Dr Herndon believes they are being produced by a nuclear reaction at the core of the Earth.

I am confident that Dr. Herndon's ideas will be proven true. If so, then you are going to see, in the next few decades, a shocking change in the science of geology, as textbooks have to be rewritten. Maybe only when scientists are shocked in that way, will they perhaps start dusting off the old Hollow Earth books, and the ancient texts and start wondering if perhaps, there is a lot more going on under our feet than they ever guessed.

I am glad Dorothy Leon has written this book. I hope it opens people's minds to other possibilities, and encourages personal spiritual and intellectual growth. In my view, if one can stimulate another's mind, then that is perhaps the most precious gift one could have imparted to that person.

Jan Lamprecht
Johannesburg, South Africa
June 2004

"Skeptics may smile and denounce our work as full of nonsense or fairy-tales. But by so doing they only justify the wisdom of the Chinese philosopher Chuang, who said that 'the things that men do know can in no way be compared, numerically speaking, to the things that are unknown' and thus they laugh only at their own ignorance." Madame H. P. Blavatsky

ABOUT THE AUTHOR: Dorothy Leon:

Born under unusual circumstances in a log cabin in the Missouri Ozarks, Dorothy has been a mystic since birth. The morning after her delivery, a wandering psychic told her mother that she had "a work to do". Conversing with angels, she was later contacted by the Ascended Master Saint Germain and other Masters of the Great White Brotherhood.

After graduating from high school and college, she studied with Walter Russel's "University of Philosophy and Science," a Kahuna Priest in Hawaii, and the "Cosmic Star Temple," where she became an ordained minister. She also belonged to an Essene group, Astara, Order of the Circle Cross, Mentalphysics, Keepers of the Flame, The Baconian Society, and Mother Mary's Circle.

She has served as a minister/counselor/writer for the past thirty years, during which she, under the direction of Saint Germain, founded the "Violet Flame Center", "New Age Center", and "Anchor of Golden Light Retreat." She also did "light work" and cleansings on the planetary ley lines, which included a trip to Peru. After years of active service, she was given a "writing ministry."

CONTENTS

ILLUSTRATIONS

CHAPTER 1
"INNER EARTH"; TRUTH OR FICTION?

THE MYSTICAL CONCEPT:

Like the old adage proclaims: "Where there's so much smoke, there must be some fire". There has certainly been a lot of smoke concerning the "Inner Earth". As stated in Ecclesiastes 1:9: "There is no new thing under the sun." The idea of an Inner Earth is not new. In addition to ancient legends, myths, and folklore, it has been written about since the printed word began. Every culture has ancient texts about it. Some of these are: *Mahabharata, Ramayana, Vedas, Puranas, Bhagavad Gita, Book of Dzyan, Egyptian Book of the Dead, Gilgamesh, Eddas, Bible, Book of Enoch, Secrets of Enoch, Popul Vuh, Emerald Tablets of Thoth, Forgotten Book of Eden, Lost Books of the Bible, Book of Jasher,* and *Book of Mormon.* The ancient people of India, Tibet, Kashmir, Atlai, China, Mongolia, Siberia, Japan, Russia, Lithuania, Hungary, France, Germany, Poland, Turkey, Holland, Britain, Ireland, Switzerland, Persia, Iran, Africa, Greece, Rome, Caucasia, the Incas, Mayas, and Aztecs, each sang or wrote about it. But how reliable are these sources and how real is this phenomena?

Madame Helena P. Blavatsky, the greatest mystic of recent centuries, calls it "Meru, the Imperishable Sacred Land, the Unreachable Continent of the Gods, Fortunate Island", or "White Island".[1] Although the initiates of the mystery schools were aware of the existence of this sacred land, it is spoken of only in "blinds". It has been held as one of the greatest esoteric secrets, "'lest it be found and desecrated by the profane". In other words, both spiritually

---

[1] *The Secret Doctrine*, H. P. Blavatsky, Theosophical University Press, CA, 1977, Vol. II, pp 6, 221, 398, 401-408.

and physically, the Inner earth is literally hidden under a great ocean of secrecy.

According to the Tibetans, the North Pole, or "White Island", cannot be destroyed by fire or water, for it is the "eternal land".[2] Concerning this forbidden subject, Blavatsky hints:

> "It is the cradle of the first man and the dwelling of the last divine mortal, chosen as the future seed of humanity. Of this mysterious and sacred land very little can be said except, perhaps, according to a poetical expression in one of the commentaries, that the 'POLE-STAR has its watchful eye upon it, from the dawn to the close of the twilight of a day of Great Breath'."[3]

This informs us that it is located at the North Pole. Blavatsky further declares:

> "Tradition says, and the records of the Great Book [*Book of Dzyan*] explain, that long before the days of Adam . . . there was an ISLAND, which for its unparalleled beauty had no rival in the world. It was inhabited by the last remnant of the race [root race][4] which preceded ours. This race

---

[2] *The Secret Doctrine,* Blavatsky, Vol. II, pp 404, 408.

[3] *The Secret Doctrine,* Blavatsky, Vol. II, pp 6, 220.

[4] A root race is a new form that is created to house the incarnating souls. Each race is governed by a Manu (Law Giver) who becomes the archetype. It is believed that there will be seven "root races" upon this planet. Many of them existed long before Adam. The Aryan people are said to be of the fifth root race.

> could live with equal ease in water, air, or fire, for it had an unlimited control over the elements. These were the 'Sons of God,' not those who saw the 'daughters of men', but the real Elohim[5] . . . who imparted Nature's most weird secrets. . . . There was no communication with the fair island by sea, but SUBTERRANEAN PASSAGES, known only to the chiefs, communicated with it in all directions."[6]

According to the ancient Puranas: "Meru is the country inhibited by beings who live ten thousand years, who are free from sickness or failing; where there is neither virtue nor vice, caste or laws, for these men are of the same nature as the gods".[7] It has been pointed out by Dean Dominic De Lucia, in his informative book, *Hollow Earth in the Puranas*, that Shamballa, the capitol city of the empire of Agharta[8]--a word meaning "ungraspable"--is in the earth's INTERIOR; that it has Vedic literature; and that Vedic teachings also exist on other planets.[9] He says the Puranas contain several other references to the "BOWELS, MIDDLE, or "HOLLOW PORTION" of the "NETHERWORLD",

---

[5] Elohim are creator gods.

[6] *The Secret Doctrine,* Blavatsky, Vol. II, pp 220-221; *Isis Unveiled,* Blavatsky. Theosophical University Press, CA, 1976, Vol I, pp 589-591. Blavatsky also mentions "the cap of the North Pole" in *The Secret Doctrine,* Vol. II, p 372.

[7] *Vishnu Purana*, Book II, Chap IV; *The Secret Doctrine,* Blavatsky, Vol. II, p 401.

[8] SHAMBALLA is also spelled SHAMBALLAH and SHAMBHALLA. AGHARTA is also spelled AGARTI, AGHARTI, OR ASGARTHA. Throughout this writing I will use Shamballa and Agharta.

[9] *Hollow Earth in the Puranas* Dean Dominic De Lucia, TGS Services, TX, 2002, pp 9. The Vedas, that are very ancient, are divided into *Pre-Vedas, Rig-Vedas, and Vedas.*

which lies near the "northern ocean".[10] Ancient Vedic texts, he says, were brought to India by the Inner-Earth Aryans.[11] These records tell of a land of the midnight sun where the sun rises and sets only once a year. Vedic Hymns speak of "The Dawn of Many Days," and "The Thirty Dawn Sisters Circling Like a Wheel". The Vedic year had only two days: the Day of the Gods and the Night of the Gods. This is an accurate description of the periods of sunlight and darkness at the poles.

According to the ancient *Book of Dzyan*:

> "In the first beginnings of life, the only dry land was on the Right end [North Pole] of the sphere, where it is MOTIONLESS. The whole earth was one vast watery desert, and the waters were tepid. . . . There, man was born on the seven zones of the immortal, the indestructible. . . . There, the gods rested, and Fohat[12] reigns ever since [hence it was the 'Garden of Eden']. . . . Thus the wise fathers say that man is born in the head of his mother [earth], and that her feet at the left end [South Pole] generated the evil winds that blow from the mouth of the lower Dragon. . . . Between the first and second [races] the eternal central land was divided by the water of life. . . . The great mother travailed

[10] *Hollow Earth in the Puranas*, De Lucia, pp 9-10. 19-35

[11] *Hollow Earth in the Puranas*, De Lucia, pp 40-41.

[12] Fohat is an electrical, fiery, virile or vital force. This great eternal energy, which may be atomic, is also referred to as "fire" or "agni", hence the residents of the inner Earth are called "Masters of Fire". Lord Edward Bulwar Lytton, in his book *The Coming Race*, called this power "VRIL".

> UNDER THE WAVES [Inner Earth] and a new land was joined to the first one which our wise men call the head-gear [the cap]. She travailed harder for the third [root race] and her waist and navel appeared above the water."[13]

Buddhist teacher, Dr. Robert E. Dickhoff, says: "Agharta is a vast UNDERGROUND terminal city, being a branch of a SUBTERRANEAN, suboceanic network of TUNNELS radiating from Antarctica. . . . Agharta exists in [UNDER] the Sangpo Valley, China and is ruled by an old Prince Mani Rimpochte. Buddhism is under his jurisdiction. The underground domain of Agharta has tunnel-delta-outlets by which it can be reached. One of them is in the Tien Shan Mountains, Tartary, Asia; another one in Afghanistan. . . . A branch tunnel ends in the swamps of the Matta Grosso, Brazil; a cave in the south-west part of the U. S. These tunnels which circle the earth were built by men of a pre-glacial age civilization flourishing on both sides of the Arctic. The builders were not of earth, but visitors, colonizers from a world now called Mars.

The Martians chose Antarctica, which was then tropical. The tunnels had several purposes: swift communication, transportation, and the exploitation of miners. They served as defensive tunnel system and underground gathering centers. They came in handy during the sinking of Lemuria and Atlantis. The people lived DEEP IN THE EARTH, BENEATH the beds of rivers, even oceans. . . . Shortly after the arrival of the Martians, came those from Venus, who resembled (reptilian) serpents and crocodiles. (Isa. 28 speaks of these 'crocodiles'.) One of them was the old 'serpent' in the 'Garden of Eden'. These Venusians stationed themselves at strategic vantage points

[13] *The Secret Doctrine,* Blavatsky, Vol. II, pp 400-401.

and caused many great disturbances. The fight between Mars and Venus for possession of earth took place many million years ago. They have now clothed themselves with human bodies."[14]

The great historian, Friedrich Max Muller, in his fifty-one volume *Sacred Books of the East*, says the "cradle of humanity was an island set in the middle of a great inland sea north of the Himalayas."[15] He also spoke of the seven "root races".

The various cultures have called the Inner Earth by many different names:

India calls it "Meru" or "White Island"

Ancient Vedics called it "Ratnasanu" or "City of the Seven Petals of Vishnu"

The Tibetans call it "Agharta", "Patala" or "Shangri La"

The Sumerians called it the "Garden of Eden", the cradle of civilization

The Romans called it "Ultima Thule"

The Scandinavians called it "Valhalla", "Land of Asar", or "Asgaard, Home of the Gods"

The Persians called it "Airyana Vaega"

The Celts called it "Duat" or "Dananda"

The Greeks called it "Hyperborea", "Elysian Fields", or "Mount Olympus"

The Egyptians called it "Land of Amenti"

The Chinese called it "Chivin, the city of a Dozen Serpents"

---

[14] *Agharta, Dr.* Robert E. Dickhoff, Bruce Humphries, Inc. Publishers, MS, 195l, pp 25-29.

[15] *Sacred Books of the East,* Professor Friedrich Max Muller, London, begun in 1875.

The Knights of the Round Table called it "Avalon"

Thomas Moore called it "Utopia" or "City of the Sun"

The Mexicans called it "Tulan"

The Aztecs called it "Maya-Pan"

It has also been called "Calcas" or "Kalki", the "Colchida" from which the Argonauts sought the "Golden Fleece"

The Inner Earth, which was evidently a tropical area before several polar shifts, was believed to be the abode of Brahma, Vishnu, Krishna, Zeus, Osiris, Shiva, Hydra, Nergal, Oannes, Poseidon, Kali, Neptune, Aphrodite, and many other gods. Meru is described geographically as PASSING THROUGH THE MIDDLE OF THE EARTH-GLOBE and protruding on either side. On its upper station are the gods of "paradise", on the nether, or South Pole, "the posts of Nergal". The South Pole of darkness has also been referred to as: "Hell, Hades, Tartarus, Shoel, Netherworld, Tuat, Abyss, or the Pit".

So, right from the beginning, there appears the concept of both positive and negative beings inhabiting the Inner Earth. Does any area ever escape the conflict between good and evil? Even "heaven" was bothered by "rebellious angels". However, Rainbow City is believed to have been built at the South Pole some two and a half million years ago to offset the negativity of the dark ones.[16]

Blavatsky says Meru connects with the central land that never perishes, the land in which the day of the mortal lasts six months and his night another six months. She says this it is not the fabulous mountain in the navel or center of the earth, but that its roots and foundations are in that navel. This so-called "World Tree" or "Yggdrasil," she says,

---

[16] *Agharta*, Dickhoff, p 59.

The World-Tree, "Yggdrasil",
by Mary Scott Christy

"extends its roots into SUBTERRANEAN regions, while its limbs reach into other planes of existence".[17]

This great ash tree, "Yggdrasil", is said to exist in "Ginnungagap", a word meaning "seemingly EMPTY" or "the great CLEFT IN SPACE". Its roots, a labyrinth of SUBTERRANEAN PASSAGES, are believed to lead to many different portions of the Inner Earth.

In the *Popul Vuh,* the Mayans spoke of a NINE-LAYERED UNDERWORLD called "Xibalba". It was populated with gods, the chief being "Itzamma". The Native Americans also speak of a NINE-layered underworld filled with strange beings. The Biblical Enoch is said to have excavated through the earth's entire NINE layers and, in each excavation, to have placed a doorway to enter the central core. The Scandinavian "World-Tree, Yggdrasil", is believed to support on its branches NINE spheres. The Egyptians symbolized this concept by the NINE stamens of the avocado tree, while the Kabala displayed NINE branches on their "Tree of Life." In each case, the tenth branch reached up to the Divine Source. Odin, founder of the priesthood of the Druids and one of the few gods who was subject to mortality, says in the *Eddas*: "No one has ever known or will ever know the vastness of the roots of that vast tree."

In addition to Dante, Plato, who has served as the interpreter of religion to all mankind, also wrote about the gods of the Inner Earth. He said the "Holy Habitation of Zeus" is located "in the CENTER OF THE WORLD, on the

---

[17] *The Secret Doctrine,* Blavatsky, Vol. II, p 401.

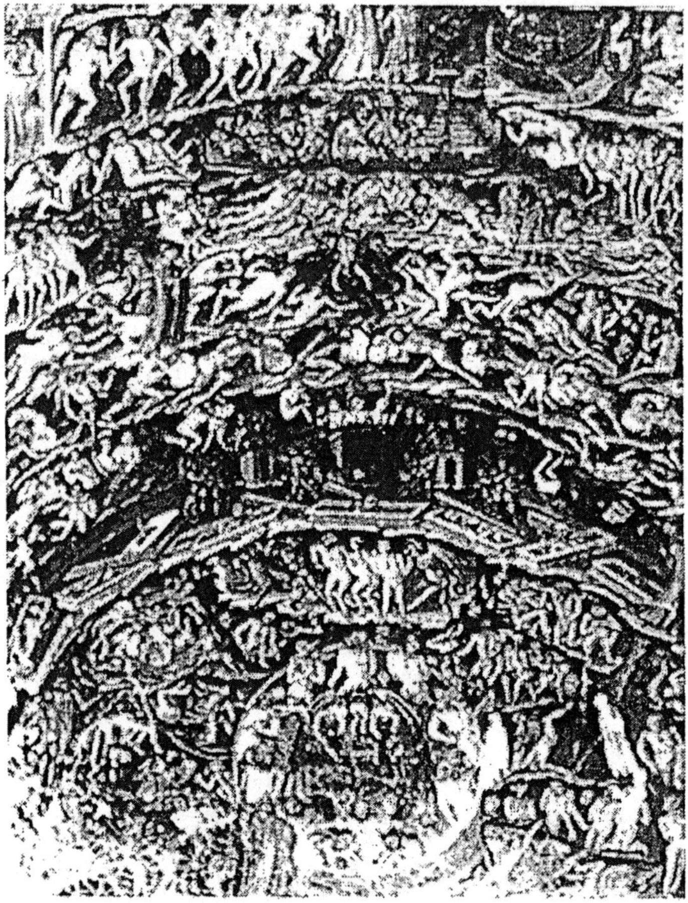

The NINE levels of the Inner Earth, Bartolomeo's view of Dante's *Inferno*

NAVEL OF THE EARTH." Plato also declared: "Apollo projects from his home among the Hyperboreans, in a land of perpetual life, BEYOND the north wind, at the EARTH'S CENTER". He also speaks of many huge CAVITIES and SUBTERRANEAN PASSAGES within the earth.[18]

The great poet, Homer, also aware of the sacred island, Hyperborea, described: "A land where the nights are short and days long, and BEYOND that land a country where the sun never sets, and the palms grow freely; a polar region where the sun never freezes".[19] What lies BEYOND the North Pole? This could only allude to a hidden region WITHIN or UNDER it, the tropical region into which Admiral Richard E. Byrd is claimed to have sailed. He said he saw cities, forests, mountains, lakes, and prehistoric animals in what he termed: "The Center of the Great Unknown".[20]

Cosmas Indicopleustes, who lived in the sixth century AD, maintained that man was born and dwelt at first in a country BEYOND the Ocean, a proof of which he claims to have received in India, from a learned Chaldean. He says: "The lands we live in are surrounded by the ocean, but BEYOND it there is another land which touches the walls of the sky, and it is this land that man was created and lived in paradise."[21] This was no doubt the biblical "Garden of Eden", the only place from which FOUR RIVERS flow, as described in the second chapter of Genesis. The Bible also mentions "the FOUR CORNERS of the earth",[22] which could

---

[18] *Critias,* Plato; *Phaedo,* Plato; *The Republic,* Plato.

[19] *The Secret Doctrine,* Blavatsky, Vol. II, pp 11-12.

[20] *A Flight to the Land Beyond the North Pole; The Missing Diary of Admiral Richard E. Byrd,* Inner Light Publications, P. O. Box 753, New Brunswick, NJ 08903.

[21] *The Secret Doctrine,* Blavatsky, Vol. II, p 399.

[22] Isa. 12:11.

only refer to the Inner Earth "Garden" where the great FOUNTAIN OF WATERS flows outward in the four directions.

The ancient Persian god, Ahura Mazda, described the Inner Earth to Zarathustra as "a place where the stars, the moon, and the sun are only once a year seen to rise and set and a year seems only as a day and night".[23] This is a clear reference to the POLAR REGIONS. The Persians elaborated:

> "Far beyond the mountains to the North, legend places the Devas and Peris [elevated ones]. Oriental tradition is ever referring to an unknown glacial, gloomy sea, and to a dark region, within which, nevertheless, are situated the FORTUNATE ISLAND, wherein bubbles, from the beginning of life on earth, the FOUNTAIN OF LIFE. The legend asserts that a portion of the first dry island, having detached itself from the main body, has remained. . . A journey of seven months' duration will bring him to that FOUNTAIN, if he keeps on journeying North straight before him as the bird flies. . . . This, of course, if one has the 'dodecpedian [meaning twelve] horse of Huschenk' or the 'WINGED Simorgh of Tahmurath' [or Taimuraz], upon which to cross over the Arctic Ocean."[24]

What are these curious "WINGED horses"? Do they refer to Pegasus, the flying horse? The wandering songsters of Persia and the Caucasus maintain a belief that far

---

[23] *The Secret Doctrine*, Blavatsky, Vol. II, p 291.

[24] *The Secret Doctrine,* Blavatsky, Vol. II, pp 398-399.

BEYOND the snow-capped summits, there is a great continent now concealed. They say it is reached by only those who can secure the services of the "twelve-legged progeny of the crocodile and the female hippopotamus, whose legs become at will twelve WINGS". It can be reached by only those who have the patience to wait for the good pleasure of "Simorg-anke", who promised that before she dies she will reveal the hidden continent to all and make it once more visible and within easy reach, by means of a BRIDGE, which the Ocean Devs will build between that portion of the "dry island".[25] The Scandinavians called this bridge "Bifrost," an appropriate name for such a frigid land.

Can you imagine a "BRIDGE" leading from the North Pole INWARD? Could it be in another dimension? The Bible says: "And an HIGHWAY shall be there, and a WAY, and it shall be called the WAY OF HOLINESS; the unclean shall not pass over it; but it shall be for those; the wayfaring man, though fools, shall not err [enter] therein."[26]

To this day the aborigines of Caucasus speak of the Holy Mountain of "Meru" as "Kap-kaz". Their bards say it requires seven months for a "SWIFT HORSE" to reach the "dry land" beyond the Kaf, holding north without ever deviating from one's way. It is interesting that Kaffeklubben is the northernmost tip of Greenland.

According to Blavatsky, the "horse" may have been a twelve-oared ship. *The Secret Doctrine* teaches that the earthly third root race built boats and flotillas before it built houses. "But," she says, "the horse though a much later animal, has nevertheless, a more occult primitive meaning".[27] The God Poseidon is, in Homer, the God of the Horse. He

---

[25] *The Secret Doctrine,* Blavatsky, Vol. II, p 399.

[26] Isa. 35:8.

[27] *The Secret Doctrine,* Blavatsky, Vol. II, p 399.

assumes that form to please Ceres. Does this indicate that he was the God of FLYING? The "twelve-legged, twelve-WINGED horse of Huschenk" was believed to be found on the Inner-Earth continent named "dry island".[28] It should become clear to the reader by now that this mysterious "horse" is some type of a "vimana" or UFO. Perhaps it is even an interdimentional "Merkabah Vehicle".[29]

So, is this phenomena of an Inner Earth, truth or fiction? The concept of an UNDERWORLD filled with UNDER PEOPLE gives the term "DOWN UNDER" an entirely new meaning. Perhaps Australia doesn't have a monopoly on the title after all. Let us continue to research the subject. Perhaps, if we DIG deep enough, we may discover it! And if we do, we may find "shining a light of such brightness, that it is no wonder the matter-of-fact skeptical scientists should be blinded by it".[30]

---

[28] *The Secret Doctrine,* Blavatsky, Vol. II, p 399.

[29] Merkabah Vehicle is a star-tetrahedron shaped spiritual, rather than a mechanical, vehicle that operates through the use of counter-rotating fields.

[30] *Isis Unveiled,* Vol. II, Blavatsky, p 548.

CHAPTER 2
DOES THE BIBLE DESCRIBE AN "INNER EARTH"?

THE BIBLICAL CONCEPT:

The previous chapter was based on various ancient texts and legends about an Inner Earth wherein dwelt various gods and root races, an area that may have been "Paradise", "The Garden of Eden," and the "Cradle of Civilization. But what does the Bible say?

> "The Almighty shall bless thee [Joseph] with blessings of heaven above, blessings OF THE DEEP THAT LIETH UNDER." Gen. 49:25
>
> "And of Joseph he said, Blessed of the Lord be his land, for the precious thing of heaven, for the dew, and for the DEEP THAT COUCHETH BENEATH." Deut. 33:13
>
> "Whatsoever the Lord pleased, that did he in heaven, and in earth, in the seas, and all DEEP places." Ps. 135:6.
>
> "Who shall descend into the DEEP." Rom. 10:7
>
> "At the name of Jesus every knee should bow, of things in heaven and things in earth, and things UNDER THE EARTH." Phlp. 2:10
>
> "Thy will be done IN EARTH [not on] as it is in heaven." Matt. 6:10
>
> "And no man in heaven, nor in earth, neither UNDER THE EARTH, was able to open the book, neither to look thereon."

Rev. 5:3

"And every creature which is in heaven, and on the earth, and UNDER THE EARTH, and such as are IN THE SEA and all that are IN THEM." Rev. 5:13

"And the angels which kept not their first estate, but left their own habitation, he hath reserved in everlasting chains UNDER darkness unto the judgement of the great day." 2 Pet. 2:4; Jude 6

"For as Jonah was three days and three nights in the whale's belly; so shall the son of man be three days and three nights IN THE HEART OF THE EARTH." Matt. 12:40

"Now that he [Jesus] ascended, what is it but that he also descended first into THE LOWER PARTS OF THE EARTH." Eph. 4:9

"And I will show wonders in heaven above, and signs IN THE EARTH BENEATH; blood and fire and vapor of smoke," Acts 2:19

"For they are all delivered unto death, to the NETHER PARTS OF THE EARTH, in the midst of the children of men, with them that go down to the pit." Ezek: 31:14, 16, 18; 31: 18, 24

What is this realm in the "NETHER, LOWER PARTS, HEART OF", or "UNDER" the earth"? The following scriptures depict a very real place with "FOUNTAINS, FOUNDATIONS, BOUNDS, DOORS, BARS, and STOREHOUSES":

"The Lord possessed me in the beginning of his way, before his works of old. I was set up from everlasting, from the beginning or ever the earth was. When there were no DEPTHS, I was brought forth; when there were no FOUNTAINS abounding with water. . . . When he prepared the heavens, I was there; when he set a compass upon the FACE OF THE DEPTH; when he established the clouds above; when he strengthened the FOUNTAINS OF THE DEEP; when he gave to the sea his decree that the WATERS should not pass his commandment; when he appointed the FOUNDATIONS of the earth." Prov. 8:22-29.

"Where wast thou when I laid FOUNDATIONS OF THE EARTH." Job 38:4

"Who hath laid the FOUNDATION of the earth, that it should not be removed forever. Thou coverest it with the DEEP as with a GARMENT. . . . Thou hast set a BOUND [IN THE WATERS] that they may not pass over; that they turn not again to cover the earth." Ps. 104:5-6, 9

"He covereth the BOTTOM OF THE SEA." Job 36:30

"For he hath founded it upon the SEAS; and established it upon the floods." Ps. 24:2

"Thou didst cleave the FOUNTAIN and the flood." Ps. 74:15

"He hath COMPASSED THE WATERS WITH BOUNDS until the day and night come to an end." Job 26:10

"Who hath BOUND the WATERS in a GARMENT?" Prov. 30:4

"It is he who calleth for the WATERS OF THE SEA, and poureth them out upon the face of the earth." Amos 5:8; 9:6

"He GATHERETH THE WATERS OF THE SEA together as an heap; he LAYETH UP THE DEPTH IN STOREHOUSES." Ps. 33:7

"Who SHUT UP THE SEA WITH DOORS, when it brake forth, as if it had issued out of the womb? . . . And brake up for it MY DECREED PLACE, and set BARS and DOORS. . . And said: Hitherto shalt thou come, but no further: and here shall thou proud WAVES BY STAYED." Job 38:8, 10-11

The phrase "my decreed place" indicates that the Inner Earth was very special. The Bible refers to it as God's CHAMBERS:

"He is in the SECRET CHAMBER." Matt. 24:26

"Who layeth the beams of HIS CHAMBERS in the waters?" Ps. 104:3

"He watereth the hills from HIS CHAMBERS." Ps. 104:13

This "secret chamber" is also called "Paradise" or the "Garden of Eden". The Bible speaks of the beautiful, towering trees in this garden, especially the cedars." Gen. 2:8; 13:10, Ezek. 31:8-9, 16, 18; 36:35. Saint Paul was "caught up into PARADISE and heard unspeakable words, which it is not lawful for a man to hear." 2 Cor. 12:4 Perhaps

Paul went to the same place that God "took" Enoch, Gen. 5:22-24; Heb. 11:5. The Bible further describes:

> "The tree of life, which is in the midst of the paradise of God." Gen. 3:24; Rev. 2:7
>
> "Like Eden, like the Garden of the Lord; joy and gladness shall be found therein, thanksgiving, and the voice of melody." Isa. 51:3
>
> "Son of man, take up a lamentation upon the king of Tyrus, and say unto him, Thus saith the Lord God; Thou sealest up the sum, full of wisdom, and perfect in beauty. Thou hast been in Eden, the Garden of God; every precious stone was thy covering, the sardius, topaz, and the diamond, the beryl, the onyx, and the jasper, the sapphire, the emerald, and the carbuncle, and gold; the workmanship of thy tabrets and of thy pipes was prepared in thee in the day that thou was created. Thou art the anointed cherub that covereth; and I have set thee so; that wast upon the Holy Mountain of God; thou hast walked up and down in the midst of the stones of fire. Thou wast perfect in thy ways from the day that thou wast created, 'till iniquity was found in thee. . . . Therefore, I will cast thee as profane out of the Mountain of God." Ezek. 28:12-16

When Adam and Eve, like king of Tyrus, were expelled from the "Garden of Eden", God set "Cherubims and a flaming sword [high-tech lasers] which turned every way, to keep the way of the tree of life." Gen. 3:23-24

Does the Bible refer to the TUNNELS,

PATHWAYS, or SUBTERRANEAN PASSAGES between the various cities or "CHAMBERS" within the "Inner Earth"?

> "Thus saith the Lord, which MAKETH A WAY IN THE SEA, and a PATH IN THE MIGHTY WATERS." Isa. 43:16
>
> "Thy WAY IS IN THE SEA, and THE PATH IN THE GREAT WATERS, and thy footsteps are not known." Ps. 77:19
>
> "Thou HATH MADE THE DEPTHS OF THE SEA A WAY [Path] for the ransomed to pass over?" Isa. 51:10
>
> "Who hath DIVIDED the watercourse?" Job 38:25
>
> "And an HIGHWAY shall be there, and a WAY, and it shall be called the WAY OF HOLINESS." Isa. 35:8

Does the Bible allude to an entrance at the North Pole?

> "He STRETCHETH OUT THE NORTH over THE EMPTY PLACE, and hangeth the earth upon nothing." Job 26:7
>
> "Out of whose womb came the ICE? And the HOARY FROST of heaven, who hath gendered it? The waters are hid as with a stone, and the FACE OF THE DEEP IS FROZEN." Job 38:29-30
>
> "Hast thou entered into the treasures of the SNOW? or hast thou seen the treasures of the HAIL, which I have reserved against the time of trouble, against the day of battle and war?" Job 38:22-23

Could this "EMPTY SPACE" be the so-called "hole in the Pole"? Could the "treasures of ICE, SNOW, FROST, and HAIL" refer to the North Pole entrance, a place where the "FACE OF THE DEEP is FROZEN"; a place that is "reserved" or protected against war? If the gods or higher root races wanted to live in an area that is protected against warring barbarians, where better to hide than inside the planet, an area that is HIDDEN by a FACE of ICE? The Bible has many references to God, or the overseer and ruler of the earth, residing in a place that is "HIDDEN".

The Bible also makes a distinction between the portions of the Inner Earth that are considered to be "the Garden of Eden", or "paradise" and those that are considered to be "hell" or the "pit". There is a great "gulf" set between the two. Perhaps this "gulf" contains the legendary "River Styx". The Bible relates:

> "And Jesus said unto him, Verily I say unto thee, Today shalt thou be with me in PARADISE." Luke 23:43
>
> "And it came to pass, that the beggar died, and was carried by the angels into Abraham's bosom: the rich man also died, and was buried: And in HELL he lift up his eyes, being in torments, and seeth Abraham afar off, and Lazarus in his bosom. And he cried and said, Father Abraham have mercy on me and send Lazarus, that he may dip the tip of his finger in water, and cool my tongue; for I am tormented in this flame. But Abraham said, Son, remember that thou in thy lifetime receiveth thy good things, and likewise Lazarus evil things; but now he is comforted and thou are torments. And besides all this, between us and you there is

A GREAT GULF FIXED; so that they which would pass from hence to you cannot; neither can they pass to us, that would come from these." Luke 16: 22-26

"By which also he [Jesus] went and preached unto the spirits in prison; which sometimes were disobedient." 1 Pet. 3:19-20

"Thou shalt be brought down to hell, to the sides of the pit." Isa. 14:15.

Going down to the CHAMBERS of death." Prov. 7:27

The First Book of Adam and Eve, one of *The Forgotten Books of Eden*,[31] states:

"And to the NORTH of the GARDEN [of Eden] there is a sea of water, clear and pure to the taste [not salty], like unto nothing else, so that through the clearness thereof, one may look into the DEPTHS of the earth." Chap. 1, verse 2

This coincides with the notes of the explores who discovered fresh water near the North Pole. They believed the fresh water came from rivers inside the Inner Earth. The icebergs in that area are also made up of fresh water. It is believed that the gravity pull around the lip of the polar openings is stronger than anywhere else on earth. It is evidently so strong that the heavy salt water remains several

[31] *The Forgotten Books of the Bible* is a compilation several ancient manuscripts. The First and Second *Book of Adam and Eve* was originally written in Arabic, then translated into Ethiopic. Dr. S. C. Malan translated it into English. It was edited by Dr. E. Trumpp, Professor at the University of Munich.

feet below the fresh water. Thus explorers could sail over the edge of the lip without being aware of it.

When Adam and Eve sinned, they were expelled from the Garden of Eden. *The First Book of Adam and Eve* says:

> "And God commanded him [Adam] to dwell there in a CAVE in the rock--The Cave of Treasures." Chap. 1, verse 9
>
> "And when they came to the opening of the gate of the Garden [of Eden], and saw the broad earth spread before them [surface of the earth], covered with stones large and small, and with sand, they feared and trembled and fell on their faces. . . . Because--whereas they had hitherto been in the garden-land, beautifully planted with all manner of trees--they now saw themselves, in a strange land, which they knew not, and had never seen. . . . Then Adam and Eve came out of the CAVE of Treasures, and drew near to the garden gate, and there they stood to look at it, and wept for having come away from it." Chap. 2, verses 2-3; Chap. 9:1

Another lost scripture, *The Book of Enoch*[32] (as translated by R. H. Charles) states:

---

[32] *The Book of Enoch* was originally written in Aramaic. Fragments of ten Enoch manuscripts were found among the Dead Sea Scrolls at Qumran. Although accepted as part of the original scriptures, it did not make it into the "Canonized Bible". A version of the original "Enoch" that survived in the Slavic language, was discovered in 1886.

"And I went from thence to the MIDDLE OF THE EARTH, and I saw a blessed place in which there were trees with branches abiding and blooming . And there I saw a holy mountain and underneath the mountain to the east there was a stream and it flowed towards the south. And I saw towards the east another mountain higher than this, and between them a deep and narrow RAVINE; in it also ran a stream underneath the mountain. And to the west thereof there was another mountain, lower than the former and of small elevation, and a RAVINE deep and dry between them; and another deep and dry RAVINE was at the extremities of the three mountains. And all the RAVINES were deep and narrow being formed of hard rock, and trees were not planted upon them. And I marveled at the rocks, and I marveled at the RAVINE, yea, I marveled very much." Chap. 26, verses 1-6 [These "ravines" agree with the Bible's description of "a great gulf".]

"The place was CLEFT as far as the ABYSS, being full of great descending columns of FIRE." Chap. 21, verse 7

"And beyond these mountains is a region the END OF THE GREAT EARTH: there the heavens were completed. And I saw a DEEP ABYSS, with columns of HEAVENLY FIRE." Chap. 18, verse 10

"And from thence I went towards the NORTH to the ENDS OF THE EARTH [North Pole], and there I saw a great and glorious device at the ends of the whole earth. And here I saw three portals of heaven open in the heaven: through each of them proceed north winds. . . . And from thence, I went to the SOUTH to the ends of the earth [South Pole], and saw there three open portals of the heaven: and thence there came dew, rain, and wind." Chap. 34, verses 1-2; Chap. 36, verses 1-2

". . . What was hidden told me what is first and last in the heaven in the height, and BENEATH THE EARTH IN THE DEPTH, and at the ends of the heaven, and on the foundation of the heaven." Chap. 60, verse 11

"And these measures shall reveal all the secrets of the DEPTHS OF THE EARTH." Chap. 61:5

"And thence I went to another place and the mountain of hard rock. And there was in it four HOLLOW PLACES, deep and wide and very SMOOTH. How SMOOTH are the HOLLOW PLACES and DEEP and DARK to look at. . . . These HOLLOW PLACES have been created for this very purpose;, that the spirits of the souls of the dead should assemble therein." Chap. 22, verses 1-4

Could the smoothness indicate its having been cut by a laser? The Inner Earth builders evidently used powerful laser beams or the "Fohat" power of "VRIL", to hew the stones, then transport and levitate them. These beams could also bore deep within the earth, through solid rock. Inner Earth explorers have also found caverns with huge stone doors leading to different passageways. A particularly interesting description of this is given in David H. Lewis' book *The Incredible Cities of the Inner Earth.*

The Book of Enoch (as translated by Richard Laurence, L.L.D) states:

> "Two [rivers] come from the north to the sea, their waters flowing into the Erythraean sea, on the east. And with respect to the remaining four, they take their course in the CAVITY OF THE NORTH." Chap. 76, verse 7 [As stated in chapter 1, the Garden of Eden is the origin of four rivers which form the "four corners" of the earth.]
>
> "And by this oath, the ABYSS has been made strong." Chap. 68, verse 27
>
> "And he concealed the spirit of Enoch in the heaven of heavens. There, I beheld, in the midst of that light, a building raised with STONES OF ICE; and in the midst of these stones, vibrations of living fire." Chap. 70, verses 6-8
>
> "From thence I afterwards passed on to another terrific place; where I beheld the operation of a GREAT FIRE; blazing and glittering, in the midst of which there was a division. Columns of FIRE struggled together to the end of the ABYSS, and DEEP WAS THEIR DESCENT." Chap.21, verses 4-5
>
> "Three separations have been made between the spirits of the dead, and thus have the spirits of the righteous been separated. Namely, by a CHASM, by WATER, and by LIGHT above it." Chap. 22, verses 9-10

You have no doubt noticed a great emphasis on the "heavenly fire". Is this the "Fohat power of VRIL", or could it be referring to an Inner Sun?

The ancient book, *The Apocalypse of Abraham*, translated from the Slavonic text, as well as the Hebrew text, states:

> "I saw the LOWER REGIONS. . . THE ABYSS. I saw the sea and its island, and its monsters and its fishes . . . I saw streams and their windings. I saw there the Garden of Eden, its flowers and flower trees and their bloom. . . . I saw a great multitude of men and women and children." Part 2, chapter 21

The Book of Mormon scriptures[33] also mentions the Inner Earth:

> "Know ye not that there are more nations than one? Know ye not that I, the Lord your God, have created all men, and that I remember those who are upon the isles of the sea; and that I rule in the heavens above and IN THE EARTH BENEATH; and I bring forth my word unto the children of men, yea, even upon ALL the nations of the earth." 2 Nephi 29:7

---

[33] *The Book of Mormon* was translated from golden plates that were found by Joseph Smith. The fourteen-year old Joseph was given a vision in 1823. In 1827, he was guided to the plates that were buried beneath a hill in the state of New York. He was able to translate them through the use of the "Urim and Thummim", instruments containing spiritual qualities. These instruments had been buried with the plates for this very purpose.

> "And behold, all things have their likeness . . . things which are in the heavens above, and things which are on the earth, and things which are IN THE EARTH, and things which are UNDER THE EARTH, both above and beneath; all things bear record of me." Moses 6:63

> "And I said unto them that it was an AWFUL GULF, which separated the wicked from the tree of life, and also from the saints of God." 1 Nephi 15:28

The Epic of Gilgamesh, an ancient Sumerian text, from which the Genesis story is believed to have been taken, also speaks of the Inner Earth:

> "At length Gilgamesh came to that great mountain whose name is Mashu, the mountain which guards the rising and setting sun. Its twin peaks are high as the wall of heaven and its paps reach down to the UNDERWORLD. At its gate the scorpions stand guard, half man and half dragon; their glory is terrifying, their stare strikes death unto men. Their shimmering halo sweeps the mountains that guard the rising sun. . . . Then he [the scorpion] called to the man Gilgamesh . . . 'For what have you travelled so far, crossing the dangerous waters?' . . . Gilgamesh said 'I must go. Open the gate of the mountain.' And the Man-Scorpion said, 'Go Gilgamesh, I permit you to pass through the mountain of Mashu and through the high ranges.' . . . Gilgamesh followed the suns road to his rising, through the mountain. . . . After nine

> leagues in the TUNNELS, he felt not the wind on his face, but the darkness was thick. . . After eleven leagues the dawn of light [INNER SUN] appeared. At the end of twelve leagues the [INNER] SUN streamed out. . . There was the Garden of the Gods."[34]

The mystery teachings have certainly concealed the knowledge of the Inner Earth, but as the Prophet Daniel predicted: "In the last days, the book will be opened and knowledge shall be increased". Dan. 12:4 We are presently living in these "last days" and, with the assistance of world-wide research, ancient writings being unearthed and translated, and the help of the Internet, little is now concealed. Like stirring a large pot of mystical soup, if you stir long enough, each little tidbit will come to the surface. So, let us keep an open mind and continue to stir.

---

[34] *The Epic of Gilgamesh*, introduction by N. K. Sanders; Penguin, MD, 1960; *Myths from Mesopotamia*, Stephanie Dalley, Oxford University Press, NY, 1989.

CHAPTER 3
THE "HOLLOW EARTH" THEORY

CONCEPTS OF RESEARCHERS, AUTHORS, AND EXPLORERS:

As indicated by the first two chapters, there is evidently a network of SUBTERRANEAN AREAS connected by a vast system of TUNNELS that branch out in all directions like the roots of a tree. Within these enormous SUBTERRANEAN TUNNELS, high-tech vehicles are said to pass both UNDER the land and UNDER the ocean. But is the Inner Earth restricted only to caverns and subterranean passages, or does it extend to a vast HOLLOW space within the CENTER. Could it resemble a mini cosmos that is large enough to contain a central sun, as well as cities, fields, forests, lakes, and even a small ocean?

Although it sounds preposterous, it is an idea that has eternally tugged on the minds of many great thinkers throughout the ages. From what we can gather from the various researchers, authors and explorers (the RAES, or RAYS, of Light), there are evidently different Inner-Earth areas. Some of the caverns are near the surface, while other areas are closer to the center of the earth. It has been estimated that there may be more actual land mass WITHIN the earth, which is mostly land, than there is on the surface, in which small continents are surrounded by immense oceans. If the earth is 8,000 miles across and the shell is only 800 miles on either side, this leaves an inner space 6,400 miles between wide. An eight-hundred-mile thick crust may be too conservative, however. Based on seismology, it may be from 1500 to 2500 miles thick. Even if the crust were 2500 miles thick, this would still leave an inner space of some 3,000 miles. With this amount of square footage, no wonder the residents are said to number in the millions.

The Hollow Globe,
from Dr. Raymond Bernard's book
***The Hollow Earth*, page 97**

According to Doreal, author of *Inner Earth* and founder of "Brotherhood of the White Temple", below the skin of the earth, lies a layer of molten rock or lava. The outer surface floats on the molten layer, causing continental drift. Below that belt, the temperatures are cooler in a vast system of caverns, some large enough to hold an entire city. Below that is a complex of tunnels, going in all directions. Below that is a HOLLOW SPHERE. At the center is a thousand-mile sphere that serves as earth's heart where life force is stored. It is the negative to the positive sun. Shamballa sits in a space warp that creates a ninth-dimensional vibration. Some of its buildings are eight million years old.[35] Some of Doreal's information has been questioned, especially his comments on continental drift.

It is fascinating to explore these ancient writings and theories, but is there any scientific evidence to strengthen the theory of an Inner or a "HOLLOW" Earth? This subject was written about by the ancient philosophers. Plato, Homer, Herodatus, and Tacitus, as well as Josephus. It was discussed in Terrien de Lacouperie's *Early History of Chinese Civilization*, Flammarion's *Astronomical Myths*, Lenormant's *Beginnings of History*, Hesoid's *Theogony*, Sayhce's *Records of the Past*, and Sir John de Maundeville's scholastic writings.

During the 6th century AD, Cosmas Indicopleustes wrote that man first dwelt in a country BEYOND or UNDER the ocean. In the thirteenth century, Dante Alighierti wrote *Inferno* about the NINE LEVELS of the Inner Earth. Interest seems to have dropped during the dark ages, but in the past five centuries, the Hollow Earth pioneers, sometimes referred to as the "Hollowists," have fostered a great deal of research and speculation. Let us examine their progress

---

[35] *The Inner Earth,* Doreal, Brotherhood of the White Temple, 7830 Oak Way, Sedalia, CO 80135, pp 1-9.

chronologically:

1570 - Martin Frobisher, who served Queen Elizabeth I, drew a map of his Arctic explorations, depicting curious CHANNELS, which were later experienced by Olaf Jansen (see *The Smokey God)*, who claimed to have visited the Inner Earth. These "LANES", reflected by the "water-sky" mirror, were also seen by other explorers,

1654, De Gofferel wrote *The Underground World* about SUBTERRANEAN CITIES

1692, British astronomer Edmund Halley, interested in magnetism and finding more than one magnetic field, proposed that the Earth was made of up "FOUR CONCENTRIC SPHERES". He also wrote articles on the Hollow Earth in the philosophical transactions of the Royal Society

1741, Swiss mathematician, Leonhard Euler, speculated that the Earth might be "hollow" and proposed an "INTERIOR SUN"

1741, Dane, Ludvig Baron von Holberg, wrote the novel, *Neil's Klim'm Journey Underground*

1751, Robert Paltock wrote *The Life and Adventures of Peter Wilkins*

1768, Alexander Colcott promoted the "HOLLOW SPHERE" idea

Early 1800's, Scottish scientist, Sir John Leslie, investigated theories on an Inner Earth and wrote *The Natural Philosophy.* He proposed that their might be two INTERIOR SUNS

18th century, Thomas Burnet wrote *Sacred Theory of the Earth*

The "Water Sky",
from William Reed's book
***Phantom of the Poles*, page 56**

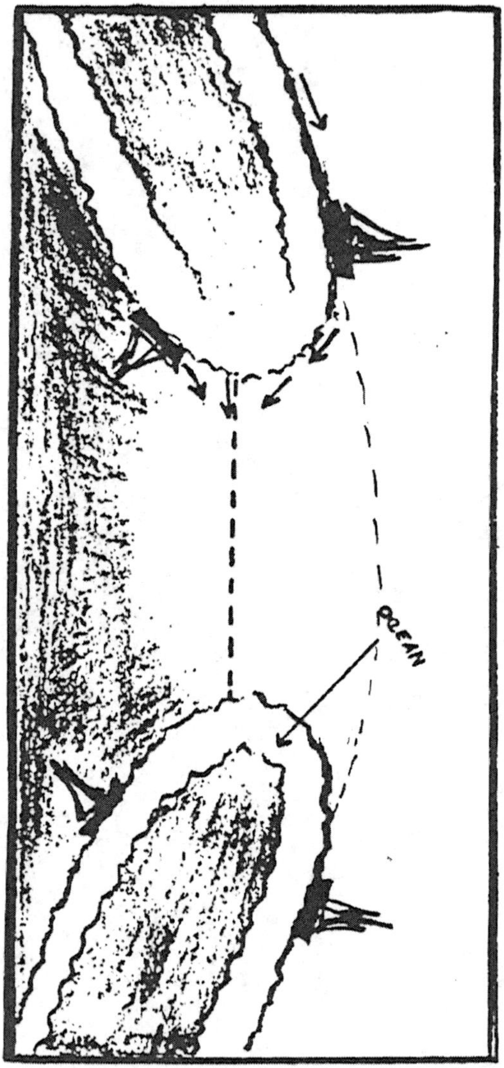

Ships sailing inside the lip of the polar opening, from Dr. Raymond Bernard's book, *The Hollow Earth*, page 98

1818, Daines Barrington wrote *On the Possibility of Approaching the North Pole*

1820, American Army Officer, John Cleves Symmes fostered Halley's "CONCENTRIC-SPHERES" theory

1826, James McBride, a millionaire, wrote *Symmes' Theory of Concentric Spheres* and petitioned congress to finance an expedition, but reached a stalemate

1827, Jeremiah Reynolds wrote a booklet, *Remarks of Symmes' Theory*. He lobbied and gave speeches in Washington D.C. to promote a polar expedition, but to no avail

1832, Capt. Adam Seaborn wrote *the novel, Symzonia: A Voyage of Discovery*, based on Symmes ideas

1838, Lt. Charles Wilkes led a four-year expedition to the South Pole. It was made possible by the fierce lobbying of Reynolds, and was the first expedition to combine a Naval crew with scientists

1838, Edgar Allen Poe wrote *The Narrative of Arthur Gordon Pym, Hans Pfaal*, and *Ms Found in a Bottle* , poems inspired by Symme's theories. He was so inspired by Reynold's speeches that he called his name as he was dying

1850, Alexander Humboldt wrote *Views of Nature: Or, Contemplations on the Sublime Phenomena of Creation* about the SUBTERRANEAN GARDENS BENEATH Cajamarca, Peru

1856, Explorer, Elisha Kent Kane, wrote *Arctic Explorations in the Years 1853-1856*, about his expedition

1864, Jules Verne, inspired by Sir John Leslie, wrote the novel *Journey to the Center of the Earth* about a CAVERN seventy-five miles BENEATH the Atlantic Ocean

1867, Explorer, Dr. Hayes, wrote *The Open Polar Sea; A Narrative of a Voyage of Discovery Towards the North Pole in the Schooner United States*

1868, Amaricus V. Symmes, son of John Symmes, wrote

The Symmes' Theory of Concentric Spheres

1870, Explorer, Capt. John Keenan and his crew, reported SEEING LAND in the Arctic. They did not realize that they had sailed over the lip of the polar opening

1873, Epes Sargent wrote *Perils and Escapes Among Icebergs; The Wonders of the Arctic World*

1877, Explorer, Julius Payer, wrote *New LANDS Within the Arctic Circle*

1879, Explorer, Adolf E. Nordenskiold, wrote *The Arctic Voyage of 1858-1878*

1879, Norwegian statesman and explorer, Dr. Fridtjof Nansen, wrote *Farthest North*, about his three-year expedition to the North Pole in his ship, "Fram". He saw what appeared to be a radiance from an "INNER SUN". He later wrote *The First Crossing of Greenland* and *In Northern Mists*

1879, Lt. George DeLong explored the Arctic in his steamer "Jeanette"

1884, J. W. Buel described the Inner Earth in his book, *The World's Wonders*

1886, Frederick Culmer wrote *The Inner World, A New Theory*

1886, Explorer, Adolphus W. Greely, wrote *Three Years of Arctic Service*

1893, L. W. Arnold wrote *The History of the Origin of all Things* in which he referred to the Hollow Earth and its formation

1894, National Geographic Magazine writer, Marcus Taker, wrote the article, *An Undiscovered LAND off the Coast of Alaska*, based on LAND sighted by the Explorers and the Eskimos who didn't realize they had entered into the lip of the polar opening

1894, Rev. Sabine Baring-Gould wrote *Curious Myths of the*

Middle Ages. He later wrote *Cliff Castles and Cave Dwellings*, describing CAVES and TUNNELS BENEATH European countries

1896, Cave explorer, Henry Mercer, wrote *The Hill Caves of Yucatan*

1898, Clement Fezandle wrote *Through the Earth*

1902, Franklin T. Ives wrote *The Hollow Earth*

1904, Dr. R. A. Harris of the U. S. Coast and Geodetic Survey, wrote in National Geographic Magazine about the possibility of there being LAND in the Arctic

1906, William Reed wrote *The Phantom of the Poles*

1907 W. J. Gordon wrote *Round About the North Pole*

1908, Willis Emerson wrote *The Smokey God*, based on Olaf Jansen and his son's ACTUAL VISIT to the Hollow Earth. Jansen claimed the benevolent beings there were fourteen feet tall

1908, Explorer, R. E. Peary, wrote *Northward*. He claimed to be the first to reach the exact center of the North Pole

1909, the "Arizona Gazette", wrote about a "SUBTERRANEAN CITY" that was found by G. E. Kincaid in the Grand Canyon

1913, Marshall H. Gardner, stimulated by the explorers findings of warmth at the poles, as well as non-Arctic wildlife, and pollen-covered snow, wrote A *Journey to the Earth's Interior or Have the Poles Really Been Discovered.* This book was so informative that it was granted a patent in 1914

1914, Lewis Spence wrote *Myths of the North American Indian. Nearly every* Indian tribe had legends about the Inner Earth

1914, Edgar Rice Burroughs, inspired by Symmes, Reed and Gardner, wrote the novel, *Pellucider - Tarzan at the Earth's Core*. Integrating modern ideas about the Hollow Earth, he also wrote about the "hollow" moon

1918, Explorer, Admiral MacMillan, wrote *Four Years in the White North*, in which he described the HILLS and VALLEYS of a LAND MASS near the North Pole. Many Arctic explorers have stated that the Eskimos of Point Barrow told them of being carried on an icecap to an ISLAND in the Arctic. Explorers, Cook and Peary, as well as Bradley, Greely, Keenan, and the Russians, later SAW LAND in the distance. They were unaware that they too had sailed around the lip of the polar opening

Explorer, Captain Richardson, wrote *The Polar Region*

1919, Harold Baylet wrote *Archaic England* about TUNNELS in Ireland and Africa

1920, Marshall H. Gardner expanded his book, *Journey to the Earth's Interior*, adding the theory of an "INNER SUN"

1922, Benjamine E. Bergenson wrote *The Earth is a Bubble*

A. Merritt wrote *The Face in the Abyss*, about exploring an UNDERGROUND CITY beneath the Peruvian Andes

1926, Amadeo Giannini wrote *Worlds Beyond the Poles,* expounding on Admiral Byrd's flight inside the North Pole. His book was expanded in 1959

1926, Pilot, Ronald Amundsen, wrote *First Crossing of the Polar Sea*, about his attempt to fly over the North Pole in a dirigible. He was carried 100 miles off course due to MAGNETIC ANOMALIES

1926, Pilot, Admiral Richard E. Byrd, made his first expedition to the Arctic; he went again in 1929, 1947 and 1956, actually FLYING INSIDE both the North and South polar opening

1928, Edna Kenton wrote *The Book of Earths* about the HOLLOW SPHERE

Around this time, the consciousness of the researchers, authors and explorers (the RAES) switched from the excitement of discovery and the concept of benevolent beings to an awareness of UFOs and the fact that were also those within the Inner Earth who were manipulative.

1929, Robert E. Howard wrote King Kull of Atlantis based on SERPENT-MEN, REPTILIANS, and ABDUCTIONS (the first of its kind). This will be further explained in chapter 9

1931, H. Spencer Lewis, wrote Lemuria: The Lost Continent of the Pacific. He later wrote The Incredible Cities of Inner Earth, a book published by Science Research Publishing House

1932, Bourke Lee wrote Death Valley Men about INNER EARTH INHABITANTS

1932, John Neihardt wrote Black Elk Speaks about Indians and their knowledge of the Inner Earth

1935, Joseph O'Neill wrote Land Under England about a totalitarian society living in caves and passageways. The book was considered by some to be an allegorical attack on Nazi Germany

1935, Michael Barton, writing as Michael X, wrote Rainbow City and the Inner Earth People, about a two-and-a-half million year old complex of seven cities that lies BENEATH Antarctica and is protected on all sides by hot springs. Barton claimed to be receiving psychic communication from the deceased Marshall H. Gardner[36]

1939, H. P. Lovecraft wrote Earth Pickman's Model about a CAVERN under Antartica. He said his writing was influenced by his "dream-ABDUCTIONS", He later wrote Lost Mines and Hidden Treasures

1942, Emery, who refused to give his last name, explored the multi-colored "rainbow" SUBTERRANEAN CITY located

---

[36] *Encyclopedia of Strange and Unexplained Physical Phenomena,* p 153.

UNDER Antarctica. (See Michael Barton or Michael X above). He said it was extremely ancient and contained the remains of a highly technological Martian civilization. Details about it were given to W.C. and Gladys Hefferlin, who wrote the A Description of Rainbow City From the Hefferlin Manuscript, a writing that is now restricted. "Rainbow City" was further described by Dr. Robert Dickhoff in 1951

In 1945 the War was over, and although writings and scientific investigations about the Inner Earth had been sparse since 1929, Ray Palmer got things rolling in full swing by writing *The Shaver Mystery* in "Amazing Stories" and "Fantastic Adventures" magazines. Richard S. Shaver, who claimed to have been TAKEN INTO the Inner Earth and to have a copy of the Lemurian alphabet, was the first person to reveal the NEGATIVE BEINGS, who he called "DEROS" (detrimental robots), who live within the Inner Earth. Palmer co-founded "Fate" magazine and later "Search", "Flying Saucers", and "The Hidden World" magazines. He also wrote *Worlds Beyond the Poles*

Ray Palmer and Richard Shaver became the father of UFOlogy. With Dr.Raymond Bernard, who also claimed to have been TAKEN INSIDE the Earth, Palmer started a newsletter. They emphasized the government's secret conspiracy. Later during the early 1950's, Palmer was warned by the Strategic Air Command to stop writing about the "hole in the pole". Their request implied that it is true and that the government had some reason for wanting the information kept top secret. The writings continued:

1946, Harold T. Wilkins wrote American Indian Myths and Mysteries

1947, Mary Douglas wrote Everglades Sea of Grass about the Inner Earth

1947, Margaret Rogers wrote Beginning about her ACTUAL VISIT to a huge CAVERN CITY where she remained for three years (details are also given in Eric Norman's 1969 book The Under People

1950, Immanuel Velikovsky wrote the very scientific Worlds in Collision

1950's, O. C. Huguenim wrote From the Subterranean World to the Sky: Flying Saucers. His concepts were based on Prof. Henrique J. de Souza, president of the Brazilian Theosophical Society and also upon Commander Paulo Strauss, Brazilian Navy Officer who believed most UFOS CAME FROM THE EARTH'S INTERIOR

1951, Buddhist teacher, Dr. Robert E. Dickhoff, wrote Agharta: The Subterranean World, in which he described "Rainbow City", an area UNDER Antarctica. This city had previously been described by Michael Barton (Michael X), Emery, and W. C. Hefferlin. Dickhoff also described the NEGATIVE "SERPENT PEOPLE" who live there

1959, Leland Lovelace wrote Lost Mines and Hidden Treasures

1960, Riley Crabb, founder and editor of "Borderland Science Newsletter", also wrote The Reality of the Underground

1968, Peter Kolosimo wrote Timeless Earth about CAVES and UNDERGROUND TUNNEL SYSTEMS

1960, Dr. Raymond Bernard wrote Agharta, The Subterranean World

1963, Dr. Raymond Bernard, who wrote The Hollow Earth, The Greatest Geographical Discovery in History, claims to have PHYSICALLY VISITED the Inner Earth through an ENTRANCE near Santa Catarina in Brazil. He later wrote Flying Saucers From the Earth's Interior

1960's Joseph H. Cater wrote The Awesome Force a book on ALTERNATIVE PHYSICS

1967, Otto Binder wrote What We Really Know About Flying Saucers

1967, Explorer, Wally Herbert, wrote Across the Top of the World. based on his polar expedition

1969, Eric Norman, who wrote The Under People, later wrote This Hollow Earth, using the name Warren Smith

1970, Bill Schmeer wrote The Entombed Miners' Staircase to Heaven about the UNUSUAL EXPERIENCES of several trapped miners

1973, Eric Von Daniken wrote The Gold of the Gods revealing the CAVE SYSTEMS BENEATH Ecuador and Peru

1974, Brinsley Le Poer Trench (Lord Clancarty) wrote Secret of the Ages: UFOs from Inside the Earth about NEGATIVE inner Earthers ABDUCTING surface dwellers. His book fostered a lot of the ideas put forth by Shaver

1975, Don Wilson wrote Our Mysterious Spaceship Moon

1975, Gary Barker wrote They Knew too Much About Flying Saucers and The Strange Case of Dr. M. I. Jessup. Shortly after reprinting Palmer's LANDS Beyond the Poles, Barker died in 1986, under very suspicious conditions

With Barker and others having died mysteriously, suddenly, the emphasis switched from a mild concern about the underground manipulators to a panic over the downright dangerous "MEN IN BLACK" (who dress in black, travel in groups of three, drive a black car, threaten severely, then disappear), and our own GOVERNMENT'S COVER-UPS and silencing tactics. What had begun as a sunny, exciting, fun exploration picnic of searching for an Inner Earth, was suddenly being drenched by a severe storm of negative encounters. Like the old adage says: If you play with fire long enough, you will eventually get burned. But the RAEs were not to be silenced. Equipped with the knowledge of UFOs, aliens, radiation belts, polar shifts, and "hole-in-the-pole" satellite photos, the Hollow Earth speculators became even more determined and organized to uncover the truth, hence:

1976, the ISCE, International Society for a Complete Earth, was founded by Danny Weiss to INVESTIGATE the Inner Earth[37]

T.H.E.R.A., The Hollow Earth Research Association, was founded

Dennis Crenshaw produced a newsletter on the Hollow Earth: The Hollow Earth Insider Research Report (T.H.E.I.)[38]

Charles A. Marcoux, who spent a quarter of a century INVESTIGATING the legends of a SUBTERRANEAN WORLD, established the Subsurface Research Center in phoenix, Arizona. He is convinced that an ENTRANCE exists in the Superstitious Mountains

Kenneth Snell founded The Hollow Earth Society of Australia and serves as editor for their newsletter

Christine Hayes founded The Rainbow Earth Dwelling Society and published "The Source" newsletter

Australia produced "Nexus" Magazine

Others dedicated RAEs continued their efforts to discover the truth:

1976, Reader's Digest Association printed an article in their book, The World's Last Mysteries, about the HOLE IN THE POLE and the EARTH'S MYSTERIOUS INTERIOR

1976, Warren Smith wrote The Hidden SECRETS of the Hollow Earth

1979, John Dyson wrote The Hot Arctic, seeking an explanation for volcanic ash on icebergs and huge stretches of melted ice ("polyanas") in the polar regions

---

[37] hotline@netcom.com

[38] onelight.com; crenshawDC@aol.com; hollow1.htmthei.html

1981, Floria Benton wrote Hollow Earth Mysteries and the Polar Shift, covering these subjects as well as GOVERNMENT COVER-UPS

1983, Bruce Walton wrote A Guide to the Inner Earth, an annotated bibliography of books and articles on the Inner Earth in other writings. He later wrote Mt. Shasta - Home of the Ancients

Other noteworthy proponents of the Hollow Earth Theory, not listed above, are archaeologist, Churchward; researchers, Evola, Godwin, and Comte de Buffon; historian, Serrano; polar explorers, Scott, McLure, Ellsworth, Nobile, and Ross. Mystic explorers, Roerich and Illion; Theosophists, Blavatsky, Steiner, Besant, and Elliot; the great seer, Edgar Cayce; and many more who foster the spiritual concept will be explained in chapter 7. All of these brave beings have forever modified the narrow ideas of mankind and pushed back the frontiers of knowledge and understanding.

There have also been several books written on the subject in recent times, many of which are mentioned or quoted in this writing. The most scientific and in-depth book I have found is: *Hollow Planets, a Feasibility Study of Possible Hollow Worlds*, by Jan Lamprecht; a real jewel. Other great books are: *Caverns, Cauldrons, and Concealed Creatures* by William Michael Mott and *World Top Secret: Our Earth is Hollow* by Rodney M. Cluff. Cluff, who believes the "Ten Lost Tribes" reside in the Inner Earth, is actually planning a polar expedition to investigate.[39] Other very good presentations are given in *Mysteries of the Inner Earth* by David Pratt;[40] *The Hollow Earth, Land of the Inner Sun,* by Rev. Ronald S. Eppich;[41] *Hollow Earth in the*

[39] http://www.ourhollowearth.com/VoyagetoHollowEarth.htm

[40] ourworld.compuserve.com/homepages

[41] http://shamans-cave.com/maxpages/The_Hollow_Earth

Puranas, by Dean De Lucia; *The Lost World of Agharti, the Mystery of Vril Power,* by Alec Maclellan; *Hollow Earth Authentic*, by Sadek Adam; and the age-old favorite, *The Hollow Earth*, by Raymond Bernard. Two excellent FICTION books that are filled with technical concepts, are: *Abduction* by Robin Cook, and *Shamballa*, by James Redfield.

CHAPTER 4
THE "HOLLOW EARTH" THEORY

THE SEMI-SCIENTIFIC CONCEPT:

The "Inner Terrans" and the "underearth cities" continue to hold a reign on man's interest. Although the scoffers have been numerous, the "hollow earthers" are still avidly pursuing the subject.

SUBTERRANEAN BASES and UNDERGROUND CITIES should not appear alien to us, since both the United States and Russia have built their own. The U. S. government constructed a huge UNDERGROUND survival city INSIDE Mount Weather in West Virginia, during the 1960's. NORAD, a defense headquarters in Cheyenne Mountain, near Colorado Springs, is another U.S. SUBTERRANEAN AREA. Many other such facilities exist. These SUBTERRANEAN AREAS are not deep within the "center" of the Earth, but fairly near the surface. They were prepared for survival areas in case of cataclysms or nuclear fallout. This is apparently why the inner earthers moved to the interior of the earth in the first place. For centuries, they have lived in their super-bunker cities hewn beneath the surface of the planet.

Why is the government COVERING UP all information concerning UFOs, polar openings, aliens, and the so-called "Crocker Land"? Why has Admiral Byrd's diary disappeared? Why have pages been removed from the diaries of Peary, Cook and Greely? (Cover-ups will be further discussed in chapter 9.)

If the earth is "hollow", how do the subterrestrials enter their realms? According to Col. Woodard:

> "The Hollow Earth residents have the ability to SPLIT THE OCEAN FLOOR and create a VORTEX, as is shown with the Bermuda Triangle. There are seven different levels in these vortexes, and equipment and beings are brought in and placed corresponding to these different levels. The vortexes act as doorways for entrance or exit to the hollow interior of the earth. There is more than one triangle area: one at Lake Erie, and another off the coast of Mexico, one off of Japan, as well as other geographic locations of the earth. These are called 'QUIET ZONES.'"[42]

This reference to "QUIET ZONES" makes one wonder if there is a relationship to the *Book of Dzyan's* reference to the sacred land being where the waters are 'MOTIONLESS."[43] However, a vortex generally indicates a spinning action.

The UFOs that are continually seen diving into the water are no doubt aware of these UNDERWATER ENTRANCES. Lakes, rivers and wells are also believed to contain such "DOORWAYS". The Inner Earth can also be reached through MINE SHAFTS, subterranean TUNNELS, and extinct volcanoes, but the most well-known PORTALS are through CAVES. Shamballa is believed to be located UNDER the area of the Tibetan Himalayas. According to Doreal, it is precisely UNDER Lhasa, Tibet about SEVENTY-FIVE MILES DEEP.[44] The heart of Agharta is

---

[42] *Traveler to the Interior of the Hollow Earth,* Part 2, p 2, Col. Billie Faye Woodard; http://onelight.com/colb/part2.htm

[43] *The Secret Doctrine,* Blavatsky, Vol. II, pp 400-401.

[44] *Shamballa or The Great White Lodge*, Doreal, Brotherhood of the White Temple, p 9.

believed to be close to Lake Manasarowar. The ENTRANCE to Patala is believed to be in the Well of Sheshna in Benares, India. Shangri-La lies BENEATH the Russian Steppes. Telos lies UNDER Mt. Shasta, and Catharia is BENEATH the Aegean Sea. UNDERGROUND CITIES are also said to exist BELOW Maui, the Himalayas, the Andes, Brazil, Ecuador, the Caroline Islands, and Iraq. Ground-penetrating radar has recently revealed a HUGE METROPOLIS UNDER the Pyramid at Giza with miles of TUNNELS and CHAMBERS.[45] Ancient CAVERN with TUNNEL SYSTEMS run throughout the world. The Eskimos even claim there is an UNDERGROUND PASSAGEWAY connecting Asia and America, running BENEATH the Bering Strait.

In a variety of writings and reports, we also find references to Inner-Earth ENTRANCES through the Mammoth Caves in Kentucky and several CAVES throughout Tennessee, Alabama, Georgia, and Missouri, the most prolific cave belt in America. DOORWAYS are also said to exist at Sulphur Springs, Colorado, within a mile of Bromide Springs; Archuleta Mesa near Dulce, New Mexico; Lechuguilla Cave in New Mexico; Carlsbad Caverns in New Mexico; Mt. Shasta, California; a circular hole in the ground near Casa Diablo north of Bishop, California; Grand Canyon, Arizona; San Francisco Mountains near Flagstaff, Arizona; near Brewster, New York; "Ghost Hole" in Connecticut; and an unnamed valley in Alaska. Lechuguilla Cave is of particular interest due to a STRONG WIND that blows from its entrance.

These GATEWAYS are said to exist throughout the world. Chiquibul Cave in Guatemala and Belize are believed to be the ENTRANCE to Xibalba. PASSAGEWAYS are also said to be in Manaus, Brazil; Roncador Mountains of

---

[45] *Hollow Earth in the Puranas*, De Lucia, pp 45-46.

Northeast Matto Grasso; at Iguasu Falls; and south of the Amazon Basin. Other ENTRANCES have been found in the Deros Caves; Yucatan, Tamup-Tocco near Cuzco; Mictolan, South America; Mt. Epomco, Italy; Malta; Black Forest, Germany; in Russia, south of Tonbou, north of Voronezia and east of Livny; Cambodia; Himalayan Mountains, Tibet; Tien Shaan Mountains, Tartary; Mongolia; Nogan Lake, Mongolia; Siberia; Mt. Soba, Japan; Gobi Desert; Rama, India; Varanais (Benares) India; Ellore, Elephanta, and the caves of Ajunta in the Chandore Range, India; Hindu Kush; Great Pyramid of Giza; Atlas Mountains, Africa; Karakorum Pass in the Altai Mountains; King Solomon's Mines; Eastern Mountain of Talishan, near Qufu Province in China, east of Karing, south of Ziling and west of Bam; Sangpo Valley, China; Tien Shan Mountains, Tartary, Asia; Afghanistan; North and South Poles; under the Atlantic Ocean; and many other areas.

If these DOORWAYS exist, why haven't more people visited the Inner Earth? Roerich says:

> "Sometimes either poisonous or vitalizing gases are spread over the Earth. . . . Sometimes you hear how the sands of the great desert shift, and for a moment disclose treasures of the entrances of subterranean kingdoms. But none would dare touch these treasures. You will hear how, in the rocks, in the most deserted mountain ranges, you can see openings which connect with these subterranean passes. . . . People have not found all things, so, let a man try to reach Shamballa without a call! You have heard about the poisonous streams which encircle the uplands. Perhaps you have even seen people dying from these gases when they come near them. Perhaps you have seen

how animals and people begin to tremble when they approach certain localities. Many people try to reach Shamballa uncalled. Some of them have disappeared forever. Only a few of them reach the holy place, and only if their karma is ready."[46]

There are, however, many accounts of people who have managed to enter into these protected areas. The most reliable is often considered to be Rear Admiral Richard E. Byrd. According to his diary:

> "During his 1947 arctic flight of 1700 miles BEYOND the North Pole, he reported by radio that he saw below him, not snow and ice, but land areas consisting of mountains, forests, green vegetation, lakes, and rivers, and in the underbrush saw a strange animal resembling a mammoth. . . . In January 1956, Admiral Byrd led another expedition to the Antarctic and there penetrated for 2300 miles beyond the South Pole. The radio announcement at this time said: 'On January 13, members of the United Stated expedition penetrated a land extent of 2,300 miles BEYOND the pole. The flight was made by Rear Admiral George Dufek of the United States Navy Air Unit.' Byrd said on March 13, 1956, 'The present expedition has opened up a vast new land.'"[47]

[46] *Shamballa*, Roerich pp 9-12, 215-216.

[47] *A Flight to the Land Beyond the North Pole: The Missing Diary of Admiral Richard E. Byrd*, Inner Light Publications, Box 753, New Brunswick, NJ 08903

Byrd said he was instructed by residents of the Inner Earth to deliver a message to the President, asking him to end all nuclear underground testing. He said that the inner eartherians viewed such testing as a threat to their network of underground cities. They feared that those living on the surface might even blow up the planet that we share. Is it merely a coincidence that the International Geophysical Year convened on Antarctica in 1957? Why did the United States and more than thirty other nations prepare unprecedented polar expeditions to penetrate the poles just after Admiral Byrd's flight?[48]

What do we actually know about the deep interior of the earth? The exact mass and density is unknown. We calculate the surface to be somewhere around 197 million square miles, and it is believed to weigh close to six trillion tons. But if the earth is a solid sphere, this amount is too small. Since the earth is a spinning mass, it seems only logical that during the its formation, while still a molten mass, the centrifugal force would fling or hurl the heavier substances outward from the axis of rotation to form the crust, thereby leaving the interior relatively hollow. If so, this would also be the case of the moon and all planets, perhaps even the Sun. This is the view presented by Jan Lamprecht.

At the exact center of rotation, there is no centrifugal force, for the gravitational force is stronger than the centrifugal. That which is left behind radiates as an if it were an interior sun. This action can be compared to the accumulation at the center of an atom, a galaxy, a comet, or a solar system.

---

[48] See: *Worlds Beyond the Poles*, Amadeo Giannini,

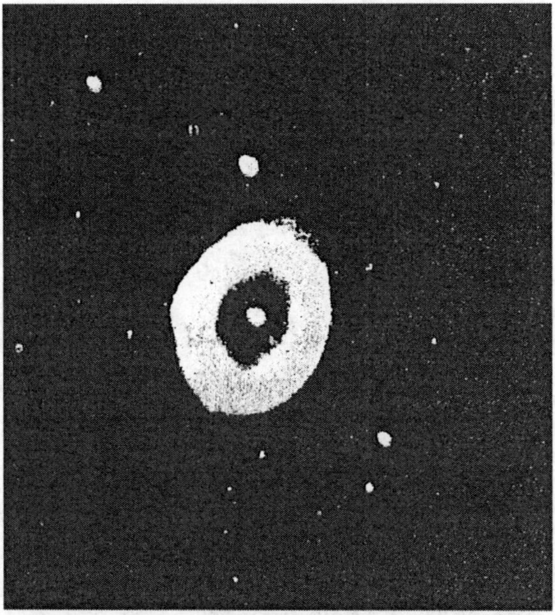

NEBULA

SPIRAL NEBULA

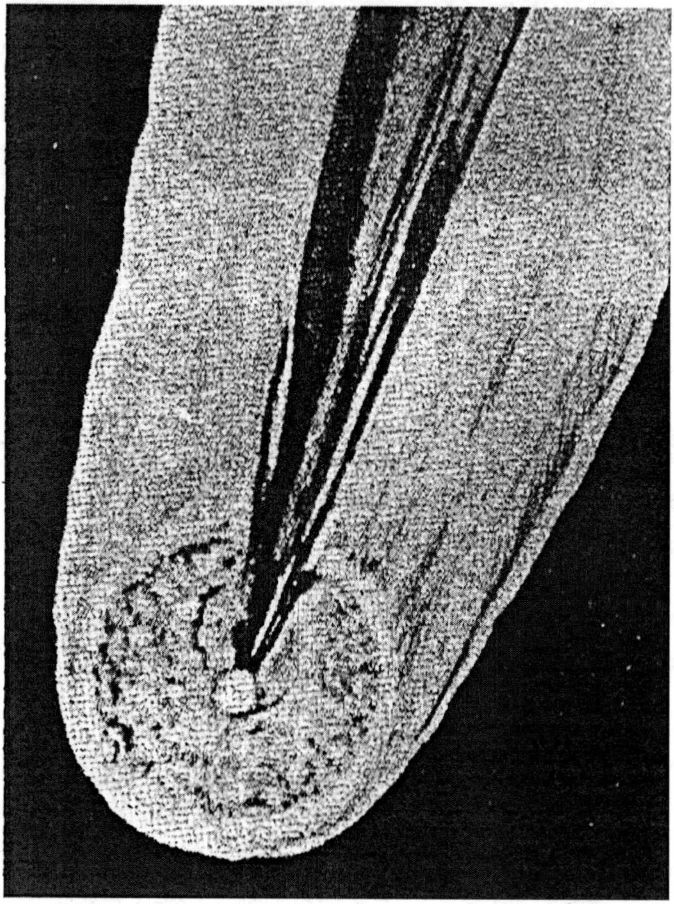

COMET
The head of Donati's Comet, 1853

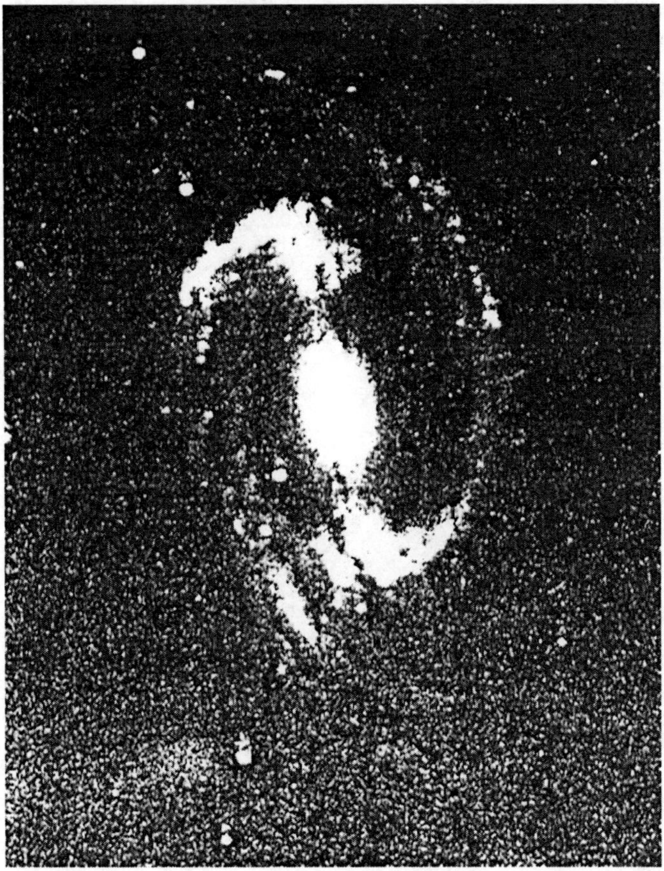

GALAXY

The crust of the earth is believed by some to be about 800 miles thick, while it is 4,000 miles to the center of the earth. The crust may, however, be even thicker. The following is a semi-scientific diagram that appeared on the cover of Dr. Raymond Bernard's book, *The Hollow Earth, The Greatest Geographical Discovery in History.*[49] As you

---

[49] This book can be ordered from Health Research, P. O. Box

can see, it depicts an inner sun at the core, as well as polar openings.

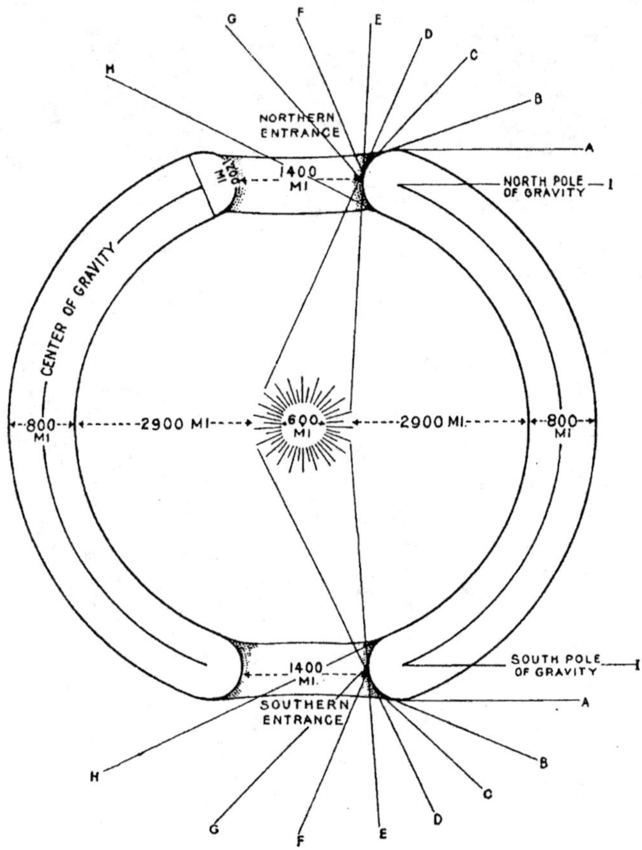

Diagram showing the earth as a hollow sphere with its polar opening and central sun. The letter at top and bottom of diagram indicate the various steps of an imaginary journey through the planet's interior. At the point marked "D" we catch our first glimpse of the corona of the central sun; at the point marked "E" we can see the central sun in its entirety. (Reproduced from "A Journey to the Earth's Interior -- or-- Have The Poles Really Been Discovered, : by Marshall B. Gardner. Printed by Eugene Smith Company, Aurora, Illinois, 1920)

850, Pomeroy, WA 99347, phone (509) 843-2385.

The following is an even more detailed version by Danish artist, Max Fyfield:

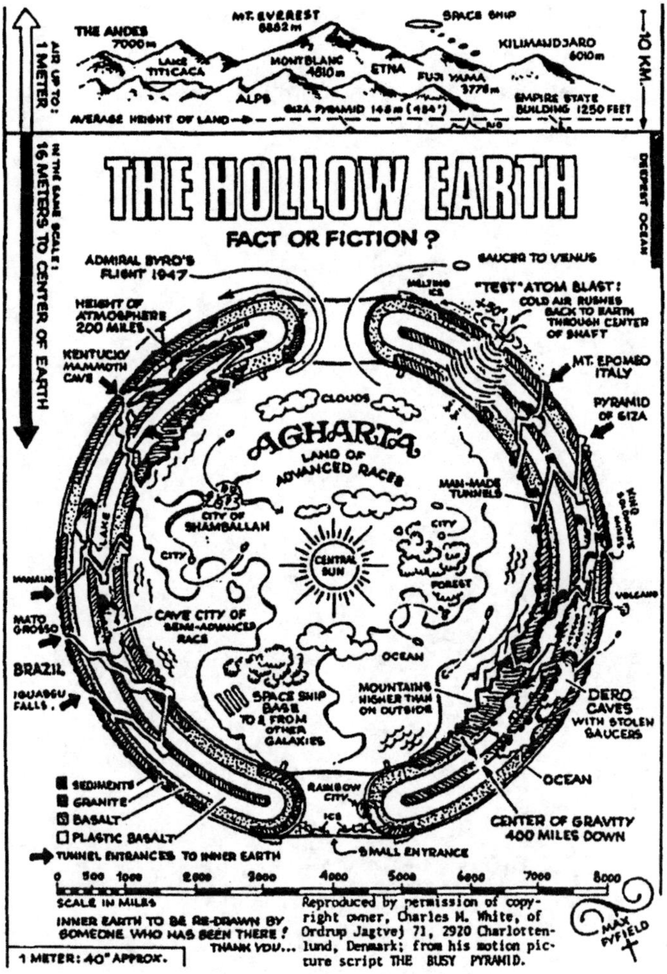

This illustration indicates catacombs, tunnels, and portals of entry.

Illustration by Mary Scott Christy

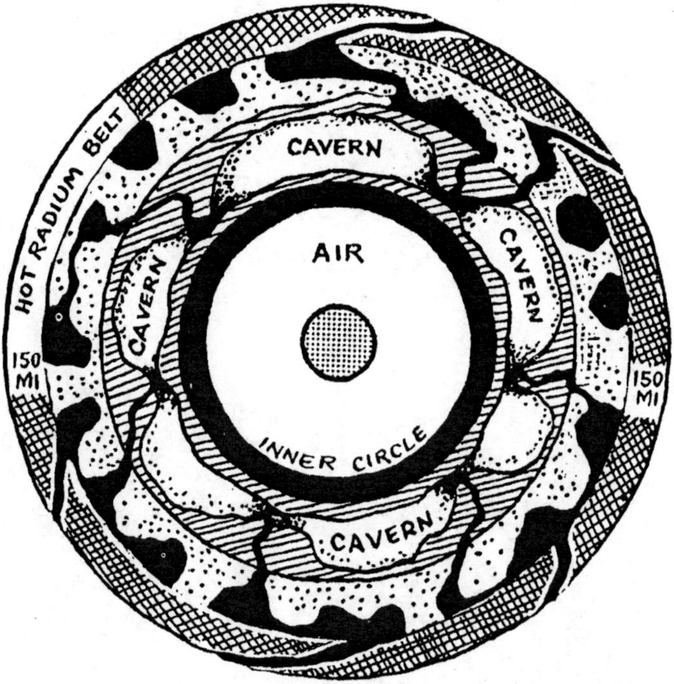

This depiction is found in Timothy Green Beckley's book, *Subterranean Worlds Inside Earth*, page 33:

As you can see, science still has a long way to go before the entire picture can be deciphered. Speculation is interesting, but now let us progress from the semi to a more scientific concept.

CHAPTER 5
THE "HOLLOW EARTH" THEORY

THE SCIENTIFIC CONCEPT:

Science News Magazine says:

> "Earth's metallic core consists of a solid iron sphere, about three-quarters of the size of the moon, sitting within an outer shell of rolling liquid iron. Hidden beneath 2600 kilometers of rock, the core has remained the most cryptic realm of our planet. Some theorists have argued that electromagnetic forces inside earth should cause the core to spin separately from the planet's outer layers. Others have hypothesized that the core should spin in synchrony with the mantle and the crust. Xiaodong Song and Paul G. Richards of the Lamont-Doherty Earth Observatory in Palisades, N.Y., have found that the inner core rotates in the same direction as the rest of the planet but about 1 degree per year faster. They report their observations in the July 18 Nature."[50]

If the inner core is moving faster than the outer surface, then every year it would logically rotate nearly one longitudinal degree further than the remainder of the earth. This discovery may help us to understand the mechanics of the earth's mysterious magnetic field. At this rate it would make a complete revolution in approximately seven hundred years. Sadek Adams informs us: "In the Middle Ages the magnetic pole was north of Siberia (A), in the sixteenth

---

[50] http:www.sciencenews.org/sn_arch/7/20_96/fobl.html

century it was east of Greenland (B). At present it's at 77°N, 101°W in Northern Canada (C)."[51] Thee following map is from Adams book, page 26:

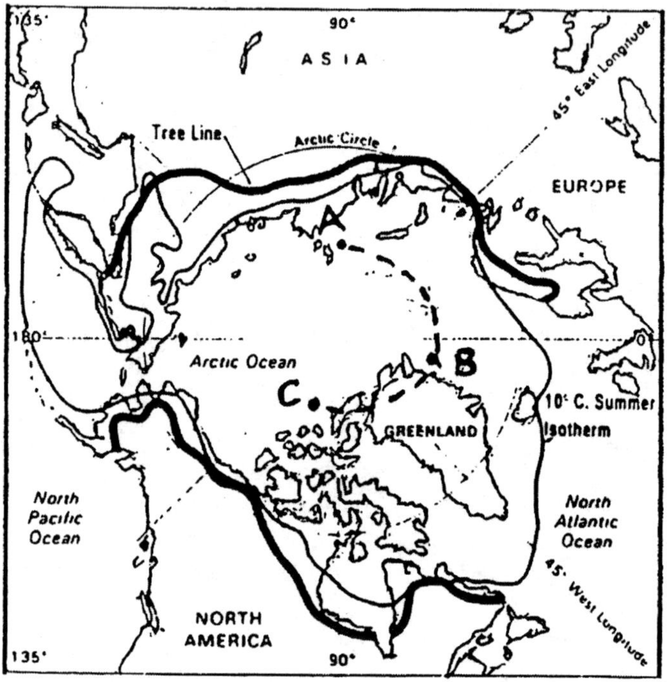

Rotation of the North Magnetic Pole

Rev. Ronald Eppich states:

"From the first recorded observation of the shift of the North Magnetic Pole in 1580 AD, the Magnetic North has shifted 2.118 miles in 235 years along the circumference to its most diametrically opposing point along the polar

[51] *Hollow Earth Authentic*, Sadek Adams, Health Research,P. O. Box 850, Pomeroy, WA 99347, 1999, p 26.

opening. That means that the North Magnetic Pole shifts along the polar opening approximately 9 miles per year."[52]

The discrepancy between the magnetic and geographic poles could, theoretically, be caused by the fact that the earth continues to tilt on its axis, while an inner sun remains stationary.

The Van Allen radiation belts, consisting of highly-charged electron and proton particles, forms a doughnut-shaped ring in the magnetic field of the earth. This belt is intense over the equator, but is absent over the areas of the polar holes. When the magnetic forcefield diminishes to an extreme degree, a polar shift can occur. Disrupted gravitational balance also causes earthquakes.

Scientists are puzzled as to why the earth's geomagnetic field does not coincide with the earth's axis, but is off center. This is seemingly clear evidence that the earth has two electromagnetic fields, one produced by an inner stationary sun, the other by the above mentioned dual rotations of the inner and outer earth. This would act as a dynamo to create an electromagnetic forcefield around the earth. If we could somehow tap the electric current at the equator, we would have an unending source of free energy. Nicola Tesla proposed this idea nearly a century ago, but few listened to him. How many are listening today? Could the energy cartels be holding them back?

It is also becoming increasingly clear that the Van Allen Radiation Belt and the auroras are not caused by solar wind, since the earth's magnetic field deflect this wind. The

---

[52] *The Hollow Earth, Land of the Inner Sun,* Rev. Ronald S. Eppich, pp 2-3

http://shamans-cave.com/maxpages/The_Hollow_Earth

energy of the solar winds do not match the intensity of the magnetoshpere. Why does the aurora shine brighter during the Arctic winter than it does in the summer? The planet Venus seems to have auroras in its atmosphere, despite the fact that it has no magnetic field. Science is, therefore, having to reexamine a lot of its pet theories. Perhaps they will someday realize the significance of the spiritual law: "As above so below; as within so without." Perhaps they will finally see that the entire universe was created in the same manner as one single atom. Atoms are hollow, consisting of an electron shell orbiting around a proton-neutron nucleus. The majority of the atom, like a solar system or the universe, consists of empty space. Perhaps each atom even has polar openings. Has anyone every investigated this possibility?

To learn more about the interior of the earth, why don't we drill into the earth to see what is there? It is approximately 4,000 miles to the center. The deepest ocean, at Marianna Trench, is 10,915 meters, or 7 miles. The deepest anyone has drilled is around 20 miles, but the earth's crustal layer is around 800 to two thousand miles thick. If the earth is 8,000 miles across and the shell is only 800 to two thousand miles on either side, this leaves an inner distance of 3,000 to 6,400 miles between each crust. It is impossible to conceive of a molten iron core and a ring of liquid iron thousands of miles in diameter.

The Veronja (or Krubera) Cave in Georgia, Abkhazia, is the deepest in the world, with a depth of 1710 meters. South African gold mines are the deepest mines, reaching depths of over 2 miles or 3.2 km. The deepest borehole drillhole ever made, near Murmansk, Russia in 1994, is around 12 km deep. At that time, scientists were surprised and embarrassed by their findings, which blew holes in many of their pet theories. For instead of temperatures increasing, after a depth of 7 km, they started to drop. A well in St. Louis, drilled to the depth of 3,843 feet, did not support the theory of internal heat either. The heat increased until they measured 3,209 feet at 107 degrees, then began to sink at 3817 to 106 degrees. At 3,827 feet, the thermometer fell to 105 degrees. At the rate, a depth of 9 miles would be below zero. This is not to say that there are not zones within the earth where very high temperatures exist, but scientific studies are beginning to prove that earthquakes, and volcanoes do not require a solid earth, with a high-temperature mantle. Lava comes from the earth's crust, approximately twenty to forty miles down.

According to William Michael Mott:

> "The crust of the earth, however, starting right below our feet and stretching down to the bottom of the Mohorovicic Discontinuity, is on average 70 kilometers, or around 43 miles in thickness. The Mohorovicic, or 'Moho' as it's called, is an anomalous region of ambiguous density that upon seismic examinations seems to be composed of hills and valleys, peaks and troughs--in other words, it is a vast, world-wide cavern system, probably varying in internal condition depending upon it's nearness to magma pockets, and lava-siphoning regions which reach down to the mantle. From this point upward, the crust is absolutely honeycombed with concavities, cracks, fissures, and the like, from the breaks between the tectonic plates and subduction zones, to aquifers reservoirs, underground rivers and streams (and ancient riverbeds which have long-since run dry), and of course caves and caverns."[53]

---

[53] *Caverns, Cauldrons, and Concealed Creatures*, William

Earthquakes generally take place at depths between 100 to 700 km, but have been known to go as deep as 450 miles below the surface. Most theories of earth's density have been based on seismic waves. But this is not an accurate source since seismic waves do not travel in straight lines. They are bent by the density and pressure as they pass through the different "honeycombed" layers of the earth. This could even cause them to curve back up to the surface. In addition to multiple reflections and refractions, there are discontinuities that also change their velocity. The layers of earth are complex, irregular, and divided by faults. P-seismic waves can travel through any substance while S-seismic waves cannot travel through gasses or liquids. Neither can travel through the hollow portion of the earth. They travel to a depth of about 1,800 miles, then change abruptly. Although a weak P-wave may occasionally reach the "shadow zone", it is due to its bending and multiple reflections between the hollow core and the surface. This could be compared to a ricocheting bullet. Thus the "shadow zone" theory is further evidence that the earth is hollow. Scientists are now reluctantly having to consider some of these formerly "outrageous" possibilities. This is depicted by the following scientific report:

> "We don't know too much about the behavior of materials at temperatures and pressure which would occur in this [interior earth] region of the earth, so the core is probably not like any liquid we encounter on the surface. Still, the fact that shear waves are not propagated through the core indicates that it is also not much like any solid we know about on the surface

---

Michael Mott, TGS-Hidden Mysteries, TX, 2002, pp 121-122.

either."[54]

The earth's crust is often confused with the lithosphere just below it. There has been much controversy about this subject. According to David Pratt:

> "The lithosphere is said to average 70 km in thickness beneath oceans, and to be 100 to 250 km thick under continents. However, seismic tomography (which produces 3D images of the earth's interior) has shown that the oldest parts of the continents have very deep roots extending to depths of 400 km or more, and that the asthenosphere is absent or very thin beneath them. Even under the oceans there is no continuous asthenosphere, only disconnected asthenospheric lenses. These facts render the large-scale movement of individual 'plates' impossible -- to say nothing of the rapid movement of the entire lithosphere!"[55]

Rodney Cluff, in his book, *World Top Secret: Our earth is Hollow*, makes an interesting observation which sheds light on the scientific dilemma:

> "Scientists need a molten asthenosphere to give them a plate-moving engine called an asthenospheric 'convection cell'. And they point to all the volcanic action at plate margins as proof of their hot

---

[54] *Physical Science 100,* Merril, Hamblin, and Thorne, Burgess Publishing Company, MN, 1978, pp 11-13.

[55] *Poleshifts, Theosophy and Science Contrasted,* Part 2: Science, Physics and Math, David Pratt, p 9.

> convection cell power-horse. What they cannot explain is how S earthquake waves travel through a liquid asthenosphere and how earthquakes occur precisely where volcanic action is uppermost. In fact, most earthquake activity occurs in the asthenosphere, so how can it be liquid? Actually, they call it 'plastic'. But plastic materials such as putty or modeling clay are not elastic. Elasticity is a prime requisite for the production of an earthquake. . . Therefore, the asthenosphere is neither plastic nor liquid but solid rock. . . . Solid rock will fracture under sudden stress."[56]

In addition to this there are still many other unknown factors such as:

The existence of antigravity

Gravity does not always increase with depth; *"Newton's Laws are Full of Flaws"*

The center of gravity is not necessarily at the center of the earth, but at a sub-sphere about two to four to seven hundred miles below the surface

The earth's magnetic field seems to rotate separately from the earth

Magnetic currents neutralize gravity

There may be sources of radiation deep within the earth shown by neutrinos (particles generated by nuclear reactions that can pass through solid matter) measured in deep mines

The continents may not be as mobile as some think

---

[56] *World Top Secret: Our Earth is Hollow*, Cluff, chapter 12, page 4;

http://www.ourhollowearth.com or e-mail: rodneyclufflcox.net

The theory of plate tectonics may be unreliable

Evidence of large submerged landmasses

Continents submerged through tectonic activity have left deep cracks and rifts known as trenches

Underground gas-chambers collapse upon themselves after losing their supporting internal gas pressure

P-wave and S-wave speeds can be inaccurate

The earth's inner core and outer layers spin at different speeds

The possibility of an anomalous curvature on the surface of the earth

Solar flares do not always trigger auroras

A mysterious blue haze that descends over the North Pole every spring

Is the earth HOLLOW? The scientific community obviously do not have all the answers.

CHAPTER 6
A "HOLE IN THE POLE"?

THE CONTROVERSIAL CONCEPT:

It is a known fact that the earth is not an exact sphere, but bulges slightly at the equator and flattens at the poles. Scientists have calculated that the earth is 16,451 miles grater in diameter at the equator than at the poles. Arctic explorers found that the ice was slightly downhill with ridges and troughs. But is it possible that the inward slanted or tilting poles could contain a "hole" that serves as an opening into the Inner Earth? In the nineteenth century, John Symmes was so certain that there was a "hole in the pole", that people began to refer to it as "Symme's hole".

This subject is still immensely controversial, with many pros, cons, and anomalies. For instance, the atmosphere is thinner at the poles, while in the majority of frozen areas it is just the opposite. What is there about the surface of the earth in that region that defies all sense of logic? It was stated in chapter 1 that the mystics claim the poles protrude, but that the protrusion is in a different dimension. This is further explained by Rev. Ronald Eppich:

> "The North and South geographic 'poles' do not really exist. They are extension of projections of the earth's curvature into space. At approximately 70 - 75 degrees north and south latitude, the earth starts to curve in. Beyond this rim, the earth flattens out and slopes gradually toward the hollow interior. In addition, the North and South magnetic poles do not coincide with the geographic poles, as they should if the earth were a solid sphere. Instead, the magnetic poles lie along the

outer rim of the magnetic circle around the polar opening, some 1400 miles in diameter, and the geographical poles lie in midair, along the projected curvature of the earth's surface."[57]

Flattening at the Polar Openings, from *Phantom of the Poles* by William Reed, p 171

How could such a "Hole in the Pole" have formed. According to Sadek Adam, author of *Hollow Earth Authentic*:

"Bodies in space are formed by

---

[57] *The Hollow Earth, Land of the Inner Sun,* Rev. Ronald S. Eppich, pp 2-3;

http://shamans-cave.com/maxpages/The_Hollow_Earth

rotating motion about a near-vertical axis and being circular or spherical has by its very nature the maximum centrifugal force at its 'equator'. where it 'bulges' most, and least at the 'poles'. The c-force at the axis extremities being weak is unable to pull material away as much as it can do at lower latitudes. So we are left with a 'hole in the pole', formed by a depression at the center of a swirl. The vortex is an empty space surrounded by rapidly rotating material, as you can observe when water goes down a bathtub drain."[58]

If there were such an opening, or "hole in the pole", why aren't our pilots seeing it? Could it be interdimensional? This question is considered by Dean Dominic De Lucia at "Skyweb":

> "This is because there is all kinds of magnetic distortion near the poles, as the magnetic lines of force pass through the inner rim of the doughnut opening. So as aircraft approach the opening, their instruments of navigation typically go haywire. At this point, it is assumed that the aircraft is basically over the pole. With the navigational instruments haywire, how is anyone going to know exactly where they are? And why would any pilot stick around to find out? The standard maneuver as a pilot passes near the magnetic distortion is to execute a right-angle turn until the navigational instruments re-orient themselves then head back in his original

[58] *Hollow Earth Authentic*, Adam, p 42.

> direction. As many times as the pilot experiences magnetic distortion, he executes right-angle turns until he works his way around the circular opening, which defines the distortion in the first place then proceeds in his original direction. In this way, even given hundreds of opportunities, aircraft only approach the rim without penetrating inwards any further."[59]

To remedy the effects of this magnetic distortion, Ray Palmer, in his magazine "Flying Saucers", during the 1950's, suggested:

> "Navigation is not to be made by the compass or by triangulation on existing maps, but solely by gyro compass on an undeviated straight course from the moment of take-off to the moment of landing. And not only a gyro compass in a horizontal plane, but one in a vertical plane also (after one enters the polar openings). There must be a positive forward motion which cannot be disputed."[60]

A vertical gyroscope would insure a pathway through the opening while a horizontal one would detect the curvature of the earth.

On the other side of the coin, we know that the many people who have visited both poles, have failed to find the proverbial "hole". In 1937, a Soviet team landed there by

---

[59] *Ring Around the Opening, Dean Dominic De Lucia.* http://skywebsite.com/hollow/arctic/id6.html

[60] Quoted in *The Hollow Earth, the Greatest Geographical Discovery in History*, Bernard, Health Research, WA, 1952, p 48.

aircraft, to set up a scientific drifting station. In 1958, the nuclear-powered submarine, Nautilus, sailed beneath the ice cap of the North Pole, from Point Barrow, Alaska, to the Greenland Sea. In 1959, the nuclear submarine Skake, did likewise. The surface vessel, Arktika, reached the North Pole from the New Siberian Islands. In 1968, Ralph Plaisted's expedition reached it by snowmobile, from Ellesmere Island. In 1969, Wally Herbert's British Expedition, reached the North Pole by dog team, travelling from Point Barrow to Spitzergen.

But did they really reach the North Pole or were their compasses fooling them? Due to several polar shifts, the poles are off to one side of the earth's axis, seemingly toward the Russian side. Explorers who have come close to the polar openings have experienced their compass needles either dipping down or pointing up, depending upon the direction from which they were approaching. Or, in some cases, the compass may even spin. Perhaps those who approached the exact spot were sucked into the polar opening to actually experience the Inner earth. This was apparently the case with Olaf Jansen and Admiral Byrd, who lived to tell about it, and the unfortunate Vikings of Greenland and the Soviet fliers, as well as those from Sweden, who flew from the Kara Sea and were never heard from again. Perhaps they are still there, reluctant to leave such a "paradise". Perhaps this was also the fate of our early satellites whose cones mysteriously disappeared as they flew at too low an altitude over the magnetic anomalies of the pole.[61] Isn't it curious that today no satellites or airline will pass directly over either pole?

In 1957, the Russians built a base in eastern Antarctica which they named "Vostok." It was later discovered that the base sat atop frozen lake Vostok. According to Richard Hoagland and Michael Bara:

---

[61] *Secret of the Ages*. LePoer T. Brinsley, p 130.

> "A team of scientists from Columbia University, working under the auspices of the NSF, early in 2001, began a series of unprecedented low-altitude aerial surveys over Lake Vostok, designed to chart gravitational, magnetic and thermal activity under the ice. In the course of doing so, they made a stunning find. A huge magnetic anomaly was discovered covering the entire Southeast portion of the shore of the Lake. This remarkable anomaly, which is discrepant from the background by over 1,000 nanoteslas (a significant variance, compared to daily variations in the earth's magnetic field), could of course be caused by 'natural' processes."[62]

The "hollow hassle" controversy rages on, but it has been said that a picture says a thousand words. The ESSA (Environmental Science Services Administration of Washington DC), took photographs of the "hole in the pole". The photo below is labeled: ESSA 7, north polar area, photo 3, taken on November 23, 1968. Shortly after these photos were released by NASA, the Government confiscated them. I saved this early photos from a spiritual newsletter that I received in 1970. Many other people did likewise, hence it has appeared in several many books about the Inner Earth.

---

[62] "What is Happening at the South Pole?", Richard Hoagland and Michael Bara;

http://www.enterprisemission.com/antarctica.htm

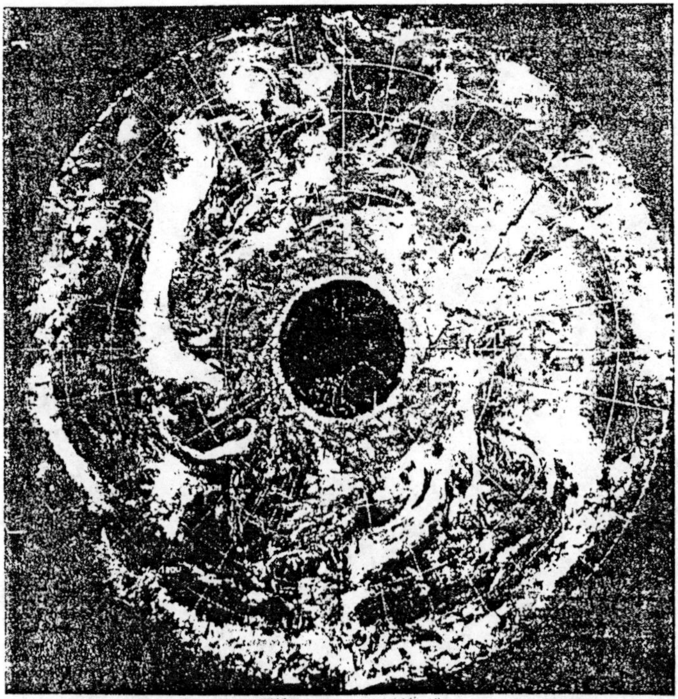

FIRST PHOTOS OF THE HOLE AT THE POLE!
Satellites ESSA - 3 and ESSA - 7 Penetrate Cloud Cover!
Mariners Also Photograph Martian Polar Opening!
(FROM FLYING SAUCERS MAGAZINE) Submitted by VELMA ACKLEY

NASA later explained that this photo was a "montage" of photos taken over a long period of time.; that the dark circle was merely the winter darkness. Since then, computer software has been programmed to handle anomalous lighting or strange spots as "errors" and to automatically "correct" them. Many photos are also touched up by a process of airbrushing. Other photos, revealing a hole in the pole have been dismissed as storms or circular air currents. But perhaps the photos speak for themselves.

The following were printed on the Internet in an

article entitled "Mysteries of the Inner Earth".[63] What do you think?

[63] http://www.marsearthconnection.com/earth.html

According to Branton:

> "One of the most important discoveries in world history was made in 1947, and most people have never heard of it today because it was only in the newspapers for one day. That year, two separate military pilots who were flying two separate aircraft on two separate missions, both saw the same incredible scene. Near the South Pole, where ice stretched to the horizon in every direction, the crews

witnessed a huge opening in the ice cap where there was a dense green forest, clear lakes, and dirt instead of snow or ice. Ice cliffs thousands of feet high surrounded this warm oasis in the ice cap. One of the pilots, Commander Bunger, radioed that the air and water temperatures in the oasis were in the mid seventies. The research crew landed a seaplane on one of the warm lakes."[64]

This discovery became known as "Bunger's Oasis".

In 1960, a newspaper in Toronto, Canada, "The Globe and Mail," published a photo of a green valley, taken by an aviator in the Arctic region.[65] Early arctic explorers claimed to have seen birds, including rose gulls, knots, swans, snow geese, blue geese, brent geese, horned-wavy geese. When Peary caught and dissected a bird, he found green grass in its craw. Otto Torell, leading a swedish expedition, sighted trees with green buds floating. Amidst the trees, they found the seed of a tropical bean. Explorer Sverdrup saw so many hares around 80 degrees north, that he named the inlet Hare Fiord.[66] Other explorers saw bees, butterflies, moths, spiders, and animals moving north as winter approached. It was presumed that they were heading to a warmer land in the north. Perhaps it was the same animals that left the Inner Earth that are found frozen in icebergs. In Siberia it is said to be crowded with mammoth bones and fossilized ivory beds. Many of the explorers said they saw musk ox, foxes, lemmings, whales, fish, mosquitos, and other wildlife. They also saw driftwood, various colors

---

[64] *Seventy Civilizations from Other Planets on Earth,* p 1; http://www.think-aboutit.com/aliens/seventy.htm.

[65] Quoted in *The Hollow Earth,* Dr. Raymond Bernard, Bell Publishing Company, NY, 1979 edition, pp 52-53.

[66] *Journey to the Earth's Interior* Gardner, pp 253-254.

of pollen and dust on the ice, as well as ice containing gravel and sand. The explorers also said they felt a warm wind blowing from the north. It is interesting that the Bible states: "He [God] causeth the VAPORS to ASCEND from THE ENDS OF THE EARTH." Ps. 135:7; Jer. 51:16

Meteorologists are finding that the Sub-Arctic is actually colder than the pole itself; that the North Pole is more than 30 degrees warmer than previously thought. The reason they give is that warm air blows to the poles from the equator. John Dyson, in his book *The Hot Arctic*, reports: "Overall there is more light in the Arctic than in the tropics. The North Pole has 140 hours more daylight than the Equator, and the Arctic Circle 230 hours more."[67] Lt. Col. C. V. Glines commented in *Polar Aviation* that there was an "inversion" and "fog" near the pole. This indicates that the warmer polar air was pushing the colder air upward. Amundsen, in 1926, and later, Soviet pilot Mikhail Gromov experienced the same thing. Some of the explorers in 1937 even claimed that it was raining at the North Pole.[68] Imagine that! But perhaps in an area where magnets go mad, auroras shine, birds migrate north, and fresh-water icebergs manifest, anything is possible, maybe even hallucinations!

The Eskimos who claim that their ancestors emerged from a fertile land beyond the ice barrier--a land where the sun never sets--have a legend about a "hole in the sea." They claim there is an open sea beyond the ice where the fishing is very good and the waters are warmer. If the poles are flattened, perhaps the sun strikes in the summer at right angles, thus shining brighter in the depression, or perhaps it is protected by a cloud mantle. caused by warmer air from the interior mixing with the colder air of the Arctic.

---

[67] *The Warm Arctic*, John Dyson, 1979.

[68] *Hollow Planets, a Feasibility Study of Possible Hollow Worlds,* Jan Lamprecht, TGS Services, TX, 1998, pp 400-401.

Rev. Ronald Eppich says:

"Around the curve of the polar openings is a ring of ice called, 'The Great Ice Barrier'. This is where icebergs form. Each winter this ring of ice is formed from the fresh water that flows out from inside the earth. During the winter months, billions of tons of free-flowing fresh water, coming from inside the earth and flowing toward the polar openings, freeze at their mouths and form mountains of fresh water ice. Since only salt water exists on the surface, fresh water ice would be impossible through any known theory. Inside these icebergs, the mammoth and other tropical plants and animals never before seen on the earth's surface exist in a state of perfect preservation. It is generally believed that these plants and animals must be from a pre-historic period of tropical environmental conditions in what is now polar regions. . . . Many animals in the far north, like the must-ox, fox, and the bear actually migrate north above the 80th parallel during the winter. If polar conditions are consistent throughout the arctic, these animals would perish. Coniferous trees wash ashore from the far north. Bees, butterflies, birds and hares exist in the far north where no vegetation should exist for their survival. Many of these species are not observed again until much further south in Alaska and Canada, where climatic conditions are much more moderate and consistent with their presence."[69]

[69] "*The Hollow Earth, Land of the Inner Sun*," Rev. Ronald Eppich ;

http://shamans-cave.com/maxpages/The_Hollow_Earth

A section of the Great Ice Barrier, from *Phantom of the Poles*, by Reed, page 225

Another interesting explanation comes from David Icke, author of "The Biggest Secret":

> "At about 70 or 75 degrees north and south latitude, the earth starts to curve

> into the polar openings to the Inner Earth, supporters of this theory suggest. It is so gradual that people who claim to have experienced this had no idea they were entering the Inner Earth until they began to see land that did not appear on the maps. These openings are an estimated 1,400 miles across and around them is a magnetic ring. The entrances are covered by clouds most of the time, advocates claim, and the airspace is restricted by law. When explorers searching for the North or South Pole reach this magnetic ring their compasses point straight down and they believe they are at the poles. They are not. They are at the magnetic ring which encircles the poles."[70]

Could this be a possible proof an a polar opening? What about the fact that the radioactive fallout from Chernobyl was concentrated in a small area near the South Pole? Doesn't this suggest some kind of an opening? And what about the swishing, rushing, crackling and hissing sounds sometimes heard during auroral displays?[71] Legends say the aurora is a reflection from an inner sun, but scientists say otherwise. A brilliant new source of thought comes from the free-thinking, yet very scientific-minded author, Jan Lamprecht, who speculates:

> "There are electric (telluric) currents which flow in the surface layer of

---

[70] *The Biggest Secret.* David Icke, Bridge of Love, MO, 1999, p 257.

[71] *Earthquakes, Tides, Unidentified Sounds and Related Phenomena,* Glen Arm, MD: Sourcebook Project, 1983, pp. 169-176; *Sizzling Skies,* New Scientist, January 6, 2001, pp. 14-19.

the earth's crust. The ground is electrically conducting and its resistivity varies markedly with depth. It has been found that these currents come from the polar regions. These currents change in sympathy with magnetic disturbances and auroras. The aurora is caused by charged particles - supposedly from the Sun. Auroras have been observed in the vicinity of nuclear tests - even at equatorial latitudes. Nuclear explosions create lots of ionized matter. So the aurora and its associated light and its ionized particles would all seem to be consistent with a nuclear reaction. I wondered if a nuclear reaction at the center of the earth could cause ionized matter to exit the earth via polar entrances, thereby helping to create the aurora - over and above the particles which come from the Sun.

What if an inner sun produced a variable stream of electrons? And what if an inner sun pulsated in some way? These streams of electrons might exit the Polar Holes, follow the magnetic lines of force and cause an aurora in the opposite hemisphere. An electron stream originating from Antarctica would therefore cause an aurora in Alaska. . . . Pulsating auroras are usually accompanied by geomagnetic pulsations, and a very rarely by faint whistling sounds. . . . Where do the ions come from which cause polar radio blackouts?

Could it be that some of the mysterious radio phenomenon on our planet are the result of our little world being hollow? . . . Many scientists have said that the earth's weather is 'manufactured' at the South Pole. Could it be that two Polar Holes in the earth's crust are the real reason why our meteorology appears to be 'chaotic'?

Why is the South Pole so important to physicists? Because here it is possible to look out into space through a kind of funnel in the earth's magnetic field, and rockets could leave the earth through these funnels."[72]

Could the South Pole be an interplanetary and intergalactic port?

From our perspective on earth, the aurora looks like a thin, waving curtain hanging in arcs from 60 to 600 miles high, but from space, it looks like a large ring surrounding the polar regions. Cluff, summarizes the aurora:

"The Aurora Borealis and Australis are caused by the solar wind of the interior sun streaming through the polar openings following the earth's electromagnetic field lines and causing atoms in the atmosphere around the openings to light up and give off the beautiful Northern and Southern 'lights'. Scientist compare the Auroras to a television set but have no answer as to what takes the place of the cathode; the energy source of the auroras is not known. . . . It is the sun within the hollow of our earth which emanates its high energy electrons

---

[72] *Hollow Planets, a Feasibility Study of Possible Hollow Worlds,* Lamprecht, Jan Lamprecht, TGS Services, TX, 1998, pp 143, 263-264, 289, 303, 313, 329, 351, 555.

> and protons through the polar openings. . . . The Van Allen radiation belts have holes at the polar extremities of the earth, coinciding with the openings there. , , , Our conclusion is that if the earth is not rigid enough to be solid all the way through and yet does not yield enough to the gravitational interactions of the moon and sun to have a liquid interior, then it must be hollow. . . . The earth trembles like a soap bubble, which is hollow. And when a large earthquake strikes, the earth vibrates like a bell, which is also hollow. The earth's fundamental period of vibration is 54 minutes. This indicates that the earth must be a hollow sphere . . . All bodies in space are hollow. Those having auroras and geomagnetic fields strongly indicate that they are not only hollow, but have interior suns and polar openings."[73]

The earth has withstood polar shifts, changes in the direction of its spin, and other cataclysmic activities. If it had been solid and rigid, it could not have withstood these events. Polar shifts are even confirmed in the Bible (Isa. 24:1), and changes in the directions of its spin is recorded in Egyptian annals.[74] Due to these upheavals, perhaps everyone is looking for the polar openings in the wrong place. Our government has a station located at the geographic South Pole and many have visited the North Pole. But the area off to one side is "RESTRICTED". Why?

---

[73] *World Top Secret: Our Earth is Hollow!* Cluff, chapter 2, pg. 25, 29.

[74] *Worlds in Collision*, Immanuel Velikovsky, Dell Publishing Company, Inc, NY, 1955, p 122.

The mystery of the "pole hole", what I call the doughnut dilemma, has not yet been solved. We need more innovative thinkers like Cluff, who is planning an expedition to explore the area at 84.4° N. Latitude and 141° E Longitude.[75] More power to him and also Danny Weiss, who is also planning an expedition.[76] Another innovative thinker, Lamprecht, said jokingly: "If one has the patience, one could try seeing if weather balloons can be sucked into the earth at one pole and then spewed out 20 months later at the other pole."[77] Do you have any suggestions?

---

[75] http://www.ourhollowearth.com/VoyagetoHollowEarth.htm

[76] Danny Weiss, P. O. Box 890, Felton, CA 95018, 1 (480) 335-9329.

[77] *Hollow Planets, a Feasibility Study of Possible Hollow Worlds,* Jan Lamprecht, TGS Services, TX, 1998, p 502

CHAPTER 7
WHO LIVES IN THE INNER EARTH?

THE GLORIOUS ONES:

Chapter 1 described the Inner Earth as home of the gods and the cradle of civilization for several different "ROOT RACES". Many books have been written about these glorious beings:

1772, according to Captain Cook's highly-classified diary found in an Army base in Texas, he met SEVEN-FOOT TALL BEINGS from Antarctica who communicated with his crew (article appeared in "Shaverton" in the summer of 1980)

18th century, astronomer and mystic, Jean-Sylvan Bailly, fostered the idea of an "ENLIGHTENED RACE" who live in the North

1885, William Warren, who believed the human race emerged from the North Pole, wrote *Paradise Found, The Cradle of the Human Race at the North Pole*

1889, Charles J. Cuteliffe-Hyne wrote *Beneath Your Very Boots*, a novel about the tall, thin, fair skinned, blue-eyed NORDIC (ARYAN) PEOPLE; later, in 1935, he wrote *My Joyful Life,* about his subterranean explorations

1893, Ingersoll Lockwood wrote *Baron Trump's Marvelous Underground Journey*

1896, John Uri Lloyd wrote the mystical *Etidorhpa* [Aphrodite spelled backwards] *Or The Ends of The Earth.* Bruce Walton later wrote that the guide in the story is believed to have been a MASTER MASON who entered through a cave near the Cumberland River in Kentucky; John H. Hart, in 1963, wrote *I Searched for Etidorhpa* about his expedition and attempt to find the SPIRITUAL "Caves of Zoroaster"

1905, Gabriel Tarde wrote *Underground Man* a novel about

the Inner Earth

1924, V. A. Obruchev wrote the novel *Plutonia* about the Inner Earth

1931, H. Spencer Lewis, author of *Lemuria: The Lost Continent of the Pacific,* revealed that remnants of a SUPER RACE from Lemuria lived in an underground city beneath Mount Shasta, California

Who else lives in the subterranean caverns of the earth? Some of the residents are evidently of an ADVANCED root race, while others have descended from other planets, producing humanoids that show evidence of genetic experiments. Some of the underworld dwellers are a MIXTURE of humans, humanoids, aliens, reptilians, serpent people, demons, djinn, fairies, trolls, satyrs, kobolds, goblins, chupacabras, elves, dwarves, hairy and scaly beings, and those who are half human and half animal. Hence, some surface dwellers, like Olaf and his son, who claim to have visited the subterranean area, say the Inner Earth beings are TALL AND BEAUTIFUL god-like specimens, while others say they have seen tiny elementals and fairies or hideous-looking beings.

There are also many tales of people who have visited the Inner Earth or have seen one of the being who surfaced, and have fallen in love with them. Merlin, the magician, was said to have been born of such a UNION. It is believed that the great beings Hercules, Achilles, Aeneas, Theseus, Plato, and Appolonius of Tyana, were also of MIXED PARENTAGE.[78] Gilgamesh was said to be TWO-THIRDS god, while Alexander, the Great, and Odin, each claimed to

---

[78] *The Secret Teachings of All Ages*, Manly P. Hall, Philosophical Research Society, CA, 1977, p cviii; *Comte De Gabalis*, Abbe N. De Montfaucon De Villars, The Brothers, London, 1670.

be HALF god. Each of them visited the Inner Earth in search of their divine ancestors, in an attempt to gain immortality. Wagner's Opera, *Der Ring de Nibelungen*, is based on the god, Odin and his adventures. Mythology informs us that Orpheus went to the Inner Earth to rescue his wife, Euridice, from Hades; Ulysses sacrificed to get the gods of the Inner Earth to respond; Pluto reigned over the underworld; and Venus was banished to dwell there. Many great heroes, such as Jupiter, Hermes, and Dionysos, are believed to have been born in caves of the underworld. It is interesting that Jesus was also born in a cave that was being used as a barn.

After emerging to live on the surface, in order to help mankind, the Agharti and Naga people strove to continue mixing their bloodlines. "Dragon blood" or "blueblood" was considered to be "royal". The pharaohs claimed to be descendants of the Inner Earth and sky gods. Some of the Inner Earthers, like the Nordics, were considered to be a "SUPER RACE".

Hitler, of course, greatly abused this information. In the 1930's Peter Bender, a German pilot, quoting from Teed and Gardner, wrote about the Hollow Earth and founded "Holt Welt Lehre", a movement that attracted several Nazi leaders. Hitler's "Nazi" group, after studying Lord Edward Bulwer Lytton's book, *The Coming Race*, about the power of "Vril", joined the "Vril Society", that was also known as the "Luminous Lodge." Hitler, the modern-day Attila, formed the "Thule Society of Bavaria" and went to Tibet and Mongolia to learn more about the Inner Earth, then dispatched an expedition to search for it. Hitler and his gang are believed to have established a base under the South Pole. Ernst Zundel, member of a Nazi Hollow Earth movement in Canada, wrote in his book, *UFOs--Nazi Secret Weapons,* that after World War II, Hitler and his last battalion escaped by submarine to their secret polar base. According to him, this is why Admiral Byrd was sent there; he was looking for Hitler.

Calling his group, "Outer Earth representatives of the 'Inner Earth', Zundel, in 1978, wrote, *Secret Nazi Polar Expeditions*, in an effort to raise funds for his own polar expedition.[79]

The Tibetans say they are descendants of a "Monkey King", and even have a genealogy book entitled: *Clear Mirror on Royal Genealogy*, to prove it. The Irish people make similar claims.

Several books and operas have been written about the UNION of man and gods, as well as beings from the Elemental Kingdom[80]:

1670, Abbe N. De Montfaucon De Villars wrote *Comte De Gabalis,* about the UNION of man and beings of the Elemental Kingdom

1730, H. Behren, M.D. wrote *The Natural History of the Harz Forest* about dwarfs who live in subterranean caverns beneath Harz

1850, Thomas Keightley wrote *The Fairy Mythology*; in 1888, he wrote *The World Guide to Gnomes, Fairies, Elves, and Other Little People*

1880, George MacDonald wrote *The Princess and the Goblin* about an underworld being MARRYING and MATING WITH a surface dweller

1886, Marie Corelli wrote *ROMANCE of Two Worlds*

---

79 *Mystic Places,* Time-Life Books, VA, pp 151-152; *Encyclopedia of Strange and Unexplained Physical Phenomena*, pp 153-164.

80 Elemental Kingdom consists of nature spirits, including gnomes, elves, leprechauns (earth); salamanders (fire); sylphs, fairies (air); and undines (water), who establish and maintain the planet. Although they manage the physical body, the planet, its water and its atmosphere, they do not create. The Elohim are the creator gods.

1896, Franz Hartmann wrote *Among the Gnomes*

1901, Robert A. Bennet wrote *Thyra: A ROMANCE of the Polar Pit*

1922, After escaping Bolshevik Russia, Dr. Ferdinand Ossendowski, fled across Mongolia, Siberia and the wilds of Asia, meeting Lamas along the way who conveyed to him many secrets. He wrote *Beasts, Men and Gods*, a book about Shamballa, the capitol city of the subterranean empire of Agharta

Dante's COMPANION, Beatrice, in *Dante's Inferno* was an Inner Earth spirit being

Wagner's opera, *Tannhauser* is about a man FALLING IN LOVE WITH a being from the Inner Earth; Tannhauser is said to remain in the mountain of Venus

Wagner's opera, *Nibelunglied* is about dwarves and the underworld

Christine Hayes, editor of *The Rainbow Earth Dwelling Society,* wrote about the Inner Earth and its different beings and creatures

This theme is also depicted in Mozart's *Magic Flute*, J.R.R. Tolekien's *Lord of the Rings* trilogy, *Alice in Wonderland, The Pied Piper of Hamlin*, *Sir Gawain and the GREEN Knight, Sleeping Beauty, Snow White and the Seven Dwarves*, *The Hunchback of Notre Dame,* and *The Little Mermaid.*

Many of these half human, half animal creatures were revered as gods, especially in Egypt and other nations. Their genetic interbreeding seemed to have given them super powers, while their memories allowed their "right of passage" back and forth from the underworld. Thus they lived in the nether regions, but occasionally came to the "overworld".

The Hopi Indians claim to have been taken into the Inner Earth for protection during cataclysmic times. They said they were cared for by the "ant people" then released through an underground opening in the floor of the Grand Canyon. The Apache Indians claim that during the cataclysm on Atlantis, they escaped by going through a tunnel system that led to a cavern called "Chicomoztoc", meaning "City of Gold". They said they lived there for a long period of time before coming back to the surface.[81] The Souix Indians were also aware of the "In-Worlders". In *Black Elk Speaks*, Neihardt says: "You shall see and have from them my power; and they shall take you to the high and lonely 'CENTER OF THE EARTH' that you may see; even to the place where the SUN CONTINUALLY SHINES, they shall take you there to understand."[82]

The Tibetans believe that the first Dalai Lama of Tibet received his instructions from Rigden-Jyepo, the "King of the World", who was believed to live in "Agharta", a great Empire within the Inner Earth. The Agharti people are said to have abandoned the surface in ancient times, due to contamination from volcanic action. Inside the earth's crust, they built a subterranean Empire, with Shamballa as its capitol. These super beings claim to have always monitored the surface dwellers and supervised our religions. Many people have seen visions or received spiritual messages from these Inner-Earth overseers. Could the Bible be referring to this "king" of the Inner Earth when it says: "Beautiful for situation, the joy of the whole earth, is Mount Zion, ON THE SIDES OF [INSIDE] the north, the city of the GREAT KING," Ps. 48:2. Perhaps the "Mount Zion" of the Israelites is the veritable 'NAVEL' of the earth, the dwelling place of

---

[81] *Caverns, Cauldrons, and Concealed Creatures*, William Michael Mott, p 30.

[82] *Black Elk Speaks* John Neihardt, University of Nebraska, NE, 1932, pp 21-22.

the "TREE OF LIFE" and the "LIVING FOUNTAINS OF WATER,". Rev. 7:17. 22:2.

In the sacred Tibetan books: *The Way to Shamballa* and *The Red Path to Shamballa*, translated by the German Oriental scholar Albett Grundwedel in 1915, Tashi, Lama III states:

> "Great Shamballa is far BEYOND the ocean. It is the mighty heavenly domain. . . . Only in some places, in the far north, can you discern the resplendent rays [aurora] of Shamballa. . . . Probably the ray from the tower of Rigden-Jyepo [king of Agharta] has reached all countries . . . In his MAGIC MIRROR he sees all events on earth. And the might of his thought penetrates into far-off lands. . . Through the Holy Scriptures we know of the teachings of the Blessed One about the inhabitants of the distant stars. From the same source we have heard of the flying steel bird . . . iron serpents which devour space with fire and smoke."[83]

It sounds like these Inner Earth beings had high-tech monitoring devices as well as a communication system with beings on other planets and solar systems. This should not surprise us since we are presently striving toward these accomplishments.

The highly-civilized Nordics, who live in Agharta, act as teachers to help mankind. It may have been these

---

[83] *Hollow Planets, A Feasibility Study of Possible Hollow Worlds,* Lamprecht, Armstrong Printing, TX, 1998, p 389; *Shamballa,* Roerich, pp 1-5

Aryan-appearing beings who brought the Vedic teachings to India. Other superior beings are called the "Tautha" or Danaans". The Dogon tribe of Mali claim their ancestors were taught by the Nommos, scaly humanoids who emerged from the sea in USOs (unidentified submarine objects). The Nomos are similar to one of the Sumerian gods, Oannes or EA, who also lived in the sea, as did Poseidon and Neptune.

One of the overseers from the Inner Earth is described by Bergier:

> "According to Peruvian legends, the demi-god Manco Capac, founder of the [Incan] empire, arose out of the depths of the earth through what the story calls 'the SPLENDID OPENING'. There were two more of these [openings] leading to NETHERWORLDS. These gates were supposed to have been located in a hill called Tampu-Tocco, some twenty miles southeast of Cuzco. . . . It is not easy to date Manco Capac . . . [he] does not appear as an ordinary historical figure; he seems to have come from outside and to be unlocated in time."[84]

There are numerous tales about "divine" beings emerging from the Inner Earth to assist the surface dwellers. The majority of religions stem from this source. Several books have been written on the subject:

18th century, Cotton Mather wrote articles on the Hollow Earth in "The Christian Philosopher"

---

[84] *Extra-terrestrial visitations from Prehistoric Times to the Present*, Jacques Bergier, Henry Regenry Company, IL, 1973, pp 37-40.

1826, Alexander Mitchel wrote *A Treatise on Natural Philosophy*

1851, William A. Bradshaw wrote *The Goddess of Atvatabar*, an occult novel with many interesting theories

1868, William Lyon wrote *The Hollow Globe*, based on information received through the clairvoyant channel, M. L. Sherman. He also wrote, *The World's Agitator and Reconciler.* A lengthy review of this book by H. P. Blavatsky is presented in Dr. Bernard's *The Hollow Earth*

1868, French Consular official, Louis Jacolliot, wrote twenty-one books, *La Bible dons l'Indie,* about ancient pre-Vedic history, revealing the mystic formula for the word AUM

1869, Cyrus Teed had a vision of the hollow or an "inverted" earth and spread his "truths" in a 300-acre utopian community in Florida, called "Koreshhanti". He wrote *The Cellular Cosmogony* in 1905 and published the magazine "Flaming Sword" until 1949

1871, Rosicrucian, Lord Edward Bulwer Lytton, wrote *The Coming Race,* a race he call the "Vril-ya" because they used the mysterious power of vril

1875, Friedrich Max Muller, who wrote the fifty-one volume *Sacred Books of the East*, dedicated them to Lord Bulwer Lytton

1884, Helena P. Blavatsky wrote Hollow-Earth related articles in the "Theosophist" Magazine. The subject was also alluded to in her books: *Isis Unveiled* and *Secret Doctrine*

1884, Theosopha and Ellora wrote *Palingenesia, or the Earth's New Birth* about the earth's future

1886, the French occultist and marquis, Saint-Yves d' Alveydre, wrote *Mission of India* about "subterranean cities" and the empire of "Agharta". He received many of his ideas while in trance, by an emissary of Agharta and from the

Dalai Lama of Tibet

1886, William and Elizabeth Denton wrote *Psychometric Researchers and Discoveries* about clairvoyant travels and explorations of the Inner Earth

1887, Hargrave Jennings wrote *The Rosicrucians, Their Rites and Mysteries*, about a stairway leading to huge caverns beneath Staffordshire

Rudolph Steiner wrote *Vril: The Power of the Coming Race*

1901, William A. Taylor wrote *Intermere* about Hollow Earth NATURE FORCES

1903, Bal Gangadhar Tilak wrote *The Arctic, Home of the Vedas*

1909, Walburgs Pagent wrote *Colloquies With an Unseen Friend* about underground descendants of Atlantis

1923. His Holiness Dalai Lama laid the cornerstone for a sacred temple in Equador." Dr. Robert E. Dickhoff says: "The country now named Ecuador shall become the spiritual center, the Tibet of the West, after Maitreya has established himself over all the Earth under the banner of Maitreya. . . . The Otovallo Indians of Ecuador have imported a High Lama from Tibet to erect a lamasery there"[85]

1924, Clark Firestone wrote *The Coasts of Illusion* reporting the discovery of an ancient Russian manuscript that described a Siberian group's journey to an UNDERGROUND CITY

1925, Alice A. Bailey, through spiritual inspiration, wrote *A Treatise on Cosmic Fire*, in which she described entities living beneath the earth's crust

1929, Harriette Curtiss wrote *Coming World Changes* about the king of the Inner Earth and his prophecies

1930, Constantine Nicholas Roerich, who visited the Himalayas and saw a flying saucer in 1927, wrote

---

[85]*Agharta,* Dickhoff, pp 38, 104.

Shamballa, which also mentions Shangri La and the empire of Agharta. He also wrote *Abode of Light*

1932, Manly P. Hall wrote *Man, the Grand Symbol of the Mysteries* which refers to the cave-dwelling Rishis of India and the sacred cavern of Zoroaster

1935, Guy W. Ballard (Godfre Ray King), founder of the "Saint Germain Foundation" and the "I AM Religious Activity, wrote, through spiritual inspiration, *Unveiled Mysteries*. The following year, he wrote *The Magic Presence*. Each of them described his visits to "subterranean areas"

1935, Eva Burton wrote *A Natural Bridge to Cross* about the degenerated race and the peaceful, intelligent beings of the Inner Earth

1939, Maurice Doreal claimed to have found and translated into English an ancient copper tablet. *The Emerald Tablets of Thoth the Atlantean* into English. It described in great detail about the building of some of the Inner Earth areas. He founded the Brotherhood of the White Temple and later wrote *The Inner Earth*

1947, Robert Charroux wrote *Treasures of Agharta*. He later wrote *The Mysterious Unknown*, *Mysteries of the Andes, Agharta and Shamballa, Treasures of Agharta,* and *The Gods Unknown*. He described an Inner Earth entrance through two Venezuelan volcanic craters. He also revealed local legends of strange beings who live in underground chambers and of UFOs seen entering and leaving the two craters

1949, Theodore Illion wrote *The Secret City in the Valley of Mystery* about an underground area beneath Tibet

1950, Dana Howard wrote *Vesta, the Earthborn Venusian* about an ancient underground civilization

1960, Theodore Fitch wrote *Our Paradise Inside the Earth* describing how it relates to biblical prophecy

1968, Bhatta Somadeva wrote *The Katha Sarit Sagara, or Ocean of The Streams of Story,* a ten volume set of East Indian lore, describing the seven subterranean worlds of Patala

1971, T. Lobsang Rampa wrote *The Cave of the Ancients* about realms under the Himalayan Mountains; in 1981, he wrote *Tibetan Sage* about the "Caves of Patala"

1982, Dr. Earlyne Chaney wrote *Revelations of Things to Come* about the Inner Earth people who descended from the stars

Rami Sajdi, of the "Order of the Labyrinth", told of a secret society that has existed for 25,000 years, during which the members built catacombs beneath the world's most sacred structures

John Godwin wrote *Occult America* about underground cities and subterranean dwellers

Are these glorious beings still appearing? The Internet folks at "Yorkshire Info", report:

> "Currently a few brave subterraneans are working on the surface. In order to blend with the masses, they have undergone temporary cellular changes so that physically they don't tower above the rest of us. They may be recognized by their gentle, sensitive nature and somewhat mysterious accent. . . . Princess Sharula Aurora Dux. the daughter of Ra and Rana Mu of Telos [an underground city beneath Mt. Shasta], has been appointed ambassador to the surface world by the Agharta Network. She is 267 years old and looks 30.
>
> The purpose of her

> Ambassadorship is to prepare the way for the merging of the two worlds; to bring the ideas, the information, and the new archetypes that will help unite our planet. Sharula has come to present a blueprint for peaceful change to those who will listen. The Aghartaens have reached a point where they cannot progress spiritually unless this merger takes place. In essence, we are one planet ascending, not half a planet. It is God's will that we take the next step together. The sooner we invite this unity, the sooner the magic will unfold. The Hierarchy has made the emergence of the subterranean cities a priority project. They are asking us to do our part by welcoming our brethren. The timing depends very much upon our receptivity and our graciousness. A successful merger is estimated within the next 10 to 20 years. . . . The gifts the Aghartaens bring are many. The secret of immortality is your birthright. The freedom to live in abundance is also your birthright.[86]

Whether or not this is true, there are ample reasons to keep an open mind. What have we got to lose? According to the above, there is much to gain. Who can scoff at a "gentle, sensitive-natured people" who can bring "peaceful change" and teach us the secrets of "immortality" and the "freedom to live in abundance"? Let us welcome them with an open heart.

But would our government welcome them? According to another channel, Aurelia Louise Jones:

---

[86] Hollowearth, pp 5 of 7
http://yorkshireufoinfo.homestead.com/hollowearthg2.html

"If they came here openly now to help us out and teach us their ways of brotherhood, they would be killed very quickly by the 'powers that be'. It has happened before. Around 1947, I am not sure of the exact date, the U.S. military had attempted to invade the city of Telos under Mt. Shasta with machine guns and their war equipment. What do you think their intentions were? The Telosians have not since been permitted by the 'Galactic Federation Council of Twelve'[87] to mix with us, surface people, and they have to abide by those rules. They used to be seen around the Mt. Shasta area at the beginning of the century but they are seen no more. It is my understanding that they make very few exceptions and for good reasons. It is really up to us if we want the benefit of their love and company among us, to grow and mature and make ourselves worthy and deserving of their presence."[88]

Perhaps being visited by beings from "the world inside the world" is not new. According to ancient Chinese, Japanese and Tibetan legends, the Naga people, the "serpent-race", who lived in the subterranean cities of Patala and

---

[87] Galactic Federation Council of Twelve is a United Federation consisting of members from Vega, Lyra, the Andromeda Constellation, Pleiades, Alpha Centauri, Tau Ceti, Epsilon Eridani, and Lumma/Ummo. They adhere to a strict nonintervention policy.

[88] *Note on the Civilizations from the Inner Earth*, Aurelia Louise Jones, ND, pp 2-3, Holistic Resource Services, Mount Shasta Light Publishing;

http://www.angelwater.net/y2k/inner_earth_beings.html; or contact Aurelia at: www.mslpublishing.com

Bhogavati, came to the surface to act as their "Dragon Kings". The archaeologists, James Churchward, who learned to decipher some of the ancient Nacaal texts, says the Holy Brotherhood of the ancient Naga is 50,000 years old. These dark-skinned people became the Ramans, who founded the seven rishi cities that acted as the centers of learning, first in Atlantis, then in the old Rama Empire of ancient India. Remnants of these teachings then spread to China, the Americas, Egypt and later to Greece and Rome. Tilak, in his book "The Arctic Home of the Vedas,"[89] states that the Vedic people's original arctic home was destroyed around 10,000 BC, the time of the sinking of Atlantis. He says that from 8,000 to 3000 BC, the Vedic people wandered before settling in India between 5,000 and 3000 BC.

The Egyptian god, Osiris, also emerged from the underworld to teach the people. Although the Nagas were highly revered and their teachings honored, the later Aryans, who came to India, considered them to be lesser beings by starting the Caste system--great shades of the Hitler consciousness! The following is a very informative channeling by Dhyana Markley:

> "The members of the 'Galactic Federation of Light' have been a part of our lives for thousands upon thousands of years. They have helped us in our individual conscious unfolding and, when necessary, have provided protection from those who would attack us. About 3,000 years ago, some off-planet raiders of the dark forces decided that they wanted to control earth and enslave all her peoples, both of Inner Earth and of the surface. Our friends of the 'Galactic Federation' helped to prevent those

---

[89] *The Arctic Home of the Vedas,* Bal Gangadhar Tilak, 1903.

of us who belong to Inner Earth from being oppressed. The dark forces continued to return until secretly and unknown to the 'Galactic Federation', they gained influence over the surface dwellers. . . . For a long time, it seemed as if the dark forces would succeed in enslaving almost all of the surface dwellers. For that reason, for a thousand years, we prepared to evacuate those who would come to us from the surface, before the chaos of the great earth changes and man's destruction of most of surface life. Fortunately for all of us, that didn't happen."[90]

Dr. Robert Dickhoff reports:

"Strange are the ways of Agharta ruled by priests and scientists, and stranger still are the mental powers involved directing the densities of those whom they choose and prepare to reach out into other dimensions in order to gather wisdom by observation even from other worlds into which these Aghartan priests dispatch their willing Lamas. . . . Arhartans should therefore alert all earth minds found free of fear, superstition, and ignorance, to unite at once and by means of spiritual communion to emit counter radiation for good from their clearing houses, Agharta, Rainbow City, and other centers in the underground

---

[90] "Conversation with Eeon of Inner Earth Regarding the Galactic Federation and Inner Earth Peoples," by Dhyana Markley, May 6, 2001 - PAO Newsletter, August 2001;

http://www.paoweb.com/newsf0801.htm

> domain, as well as in Lamaseries, Ashrams, Pagodas above the surface on earth, to combat the efforts bent on creating destructive weapons deadly to mankind."[91]

The general consensus of all who have visited or written about Agharta, it that it is a paradise; a Utopia. It is a place that is brightly lighted, and where music is played continually to keep the residents healthy and harmonious. Healing, art, crafts, culture and spirituality are encouraged. Full spectrum lighting is achieved through a process of energizing stones. They have discovered ways in which only seven acres can feed a half a million people, who are all strict vegetarians.

This light within the Inner Earth was described by Lord Edward Bulwer Lytton as "a diffused atmospheric light, soft and silvery."[92] Dr. Dickhoff described: "Caverns illuminated by a green luminescence which aids underground plant life there and lengthens human life."[93] Olaf Jansen, in *The Smokey God*, spoke of: "a circular shape with a hazy-red, bronzed appearance, which would change to a white like a luminous cloud; a smoky furnace-colored sun."

But like all realms, whether it be on Earth, within the Earth, in outer space, upon any other planet or galaxy, even within the angelic kingdom, one will find a mixture of both positive and negative beings. The inner Earth is apparently no exception, for the ancient legends and texts speak of the benevolent beings mentioned above as well as the hideous-looking violent beings who also dwell within the cities and caverns of darkness.

---

[91] *Agharta*, Dickhoff, 50, 89-90.

[92] *The Coming Race* Lytton, p 10.

[93] *Agharta* Dickhoff, p 49.

The divine ones evidently allow good and evil to co-exist that we might learn to decipher between them and choose to align with that which is good. Perhaps a good challenge increases our consciousness. In the Biblical Book of Job, Satan even appeared before God to obtain permission to test him, Job 1:6-12.

CHAPTER 8
WHO LIVES IN THE INNER EARTH?

THE INFAMOUS ONES:

What about the NEGATIVE inhabitants of the "within world"? As mentioned in chapter 7, there are rampant reports of aliens, "lizard people", serpents, and all types of hideous-looking beings who have embedded themselves into some of the Inner Earth caverns. They are believed to have mated with other creatures as well as with humans to create very strange-looking beings. Could the "cauldrons" of so-called witches, have been test tubes for mixing genetic materials. Some of the underground creatures and beings are believed to have eaten the human and animal sacrifices. They are said to have fed on the energy of fear and adrenalin Since these practices are no longer performed, the creatures are presently believed to be ABDUCTING both children and animals.

Much has been written about these "creatures" of the Inner Earth. Sadek Adam calls them a "motley cocktail of denizens".[94] The god, Odin, had to fight the "frost giants", the "hounds on the barrier", of the underworld in order to gain entrance.

Several strange creatures, sighted around the cave-ridden areas of Kentucky, Tennessee, Alabama and West Virginia, have been dubbed by various titles such as: the "Hopkinsville Goblins, Dover Demon, Jersey Devil", and "Mothman". In South America, very hideous-looking beings are called "Chupacabras". The Bible speaks of "crocodiles, leviathans, fiery flying serpents, and giants".

Various books have been written about these

---

[94] *Hollow Earth Authentic,* Adam, p 81.

NEGATIVE creatures:

St. Patrick wrote *The Colloquy of the Ancients* about encounters with the negative gods of the underworld

1918, T. Write wrote *Saint Patrick's Purgatory: A Medieval Pilgrimage In Ireland* tells of St. Patrick's journeys into the lower world near Station Island and his terrifying experiences

1960, Howard R. Patch wrote *The Other World, According to Descriptions in Medieval Literature* about St. Patrick's adventures in the "Caves of Purgatory"

These "hellish" beings nearly always appear at nights to do their ABDUCTING and to gather their genetic material from humans, plants and animals. *W*hy do they live underground and only come out at night? Perhaps they have a low tolerance for the ultraviolet radiation in sunlight, or perhaps there was a change in the sun that began to affect their genetic structure. Thus they need to mate with humans to improve their genes, or in some cases, to gain a soul.

The "little greys", who abduct humans by beaming them up into their spacecrafts, are trying to create a "super-race". They operate on the premise that their superior intellectual and intuitive capacities blended with the human's stronger body and ability to feel, will create a super being. An excellent book on this is: *Abduction, Human Encounters with Aliens,* by Pulitzer Prize winner John E. Mack, M.D., a psychiatrist who placed the abductees under hypno-therapy to stimulate their memories of alien encounters.[95]

Some beings, due to their electromagnetic makeup, also had a sensitivity to iron. In 1900, John Gregorson Campbell, wrote *Superstitions of the Highlands and Islands*

---

[95] *Abduction, Human Encounters with Aliens*, John E. Mack, M.D., Charles Scribner's Sons, Macmillan Publishing Co., NY

of Scotland, warning the reader to carry a piece of iron as a "protection against Fairies". But before getting overly involved in the subterranean circus and their manipulations, let us examine the Biblical reports of who lives there:

> "Dead things are formed from UNDER THE WATERS, and the INHABITANTS thereof." Job 26:5
>
> "Who shall DESCEND INTO THE DEEP?" Rom. 10:7
>
> "The WATERS SAW THEE, O God, the waters saw thee; THEY WERE AFRAID; the DEPTHS ALSO WERE TROUBLED." Ps. 77:16

What else does the Bible say about this inner realm? Perhaps we can find a clue in the word "WHALE," since it was a "WHALE" that swallowed Jonah. The Bible explains:

> "Now the Lord had PREPARED A GREAT FISH to swallow up Jonah." Jonah 1:17
>
> "Am I a SEA or a WHALE that thou settest a watch over me?" Job 7:12

Could this "whale" be some sort of a UFO or a USO (unidentified submarine object) that is capable of ABDUCTING a person and taking him into the "HEART OF THE EARTH"? Modern submarines have pursued these mysterious USOs. There have also been countless reports of ABDUCTED fishermen. One of the most notorious underwater sightings was at Shag Harbor. Both the United States and Canadian Navy pursued it for days.

The Biblical footnotes for Ps. 74:13 and Ezek. 32:2 say the Hebrew word WHALE is interchangeable with

DRAGON. The Bible details:

> "Art thou not it that hath cut RAHAB and wounded the DRAGON? Art thou not it which hath dried the sea, the WATERS OF THE GREAT DEEP?" Isa. 51:9-10
>
> "Thou didst divide the sea by thy strength; thou BREAKEST THE HEADS OF THE DRAGONS IN THE WATERS; Thou BREAKEST THE HEADS OF LEVIATHAN in pieces." Ps. 74:13-14
>
> "In that day the Lord with his sore and great and strong sword shall punish LEVIATHAN, the piercing SERPENT, even LEVIATHAN THAT CROOKED SERPENT; and he shall slay THE DRAGON THAT IS IN THE SEA." Isa. 27:1

The plot thickens. Now, in addition to the whale or dragon, we have two new characters: RAHAB and LEVIATHAN, the crooked serpent. What does this represent? The "CROOKED SERPENT" is the constellation Draco, the root word for DRAGON. Some of the alien "Dracos" are also believed to live in the Inner Earth. The extraterrestrial, RAHAB, was the sky-beast TIAMAT that plowed through the solar system, first breaking the Planet Maldek into the Asteroid Belt (what the earlier versions of the Bible referred to as the "hammered-out bracelet"), then plunging into the Pacific Ocean.

According to Blavatsky;

> "The DRAGON is the male principle [while] TIAMAT [or RAHAB], the embodiment of the Spirit of Chaos, of the

> DEEP, or ABYSS, is the female WATER principle, the womb. . . . The symbol of the fish originated with Vishnu. . . . The 'big fish' of Jonah is CETUS, the latinized form of KETO, and KETO is DAGON, OANNES, or POSEIDON, the WATER GOD who is half man and half FISH."[96]

What an adventure -- a "fish god" that lives in the "deep," a "whale" that ABDUCTS people, and a "Lord" that "cuts, wounds, and breaks the head of the dragon"! As stated in chapter one, perhaps the "gods" live close to the North Pole, while the "dragons" live near the South Pole. Could the South Pole be Revelation's "BOTTOMLESS PIT," ruled by "Abaddon"; a place where God cast the fallen angels and delivered them "into chains of darkness."[97] Is this also the place where the "beast ROSE UP OUT OF THE SEAS"?[98] Are these the "gates of hell," the "shoel", "tartarus", or "hades" wherein some of the dead are held?[99] Are these beings the "spirits in prison" to whom Jesus went to preach?[100]

But evidently not all of the "fallen ones" or "dragons" were "chained", for many are still allowed to roam the Earth, both the inner and the surface. The Bible states:

> "O Lord, how manifold are thy works! In wisdom hast THOU MADE THEM ALL; the earth is full of thy riches. So is this GREAT AND WIDE SEA, wherein all

---

[96] *Isis Unveiled*, Blavatsky, Vol II, pp 257-258; *The Secret Doctrine*, Blavatsky, Vol. II, p 104.

[97] Luke 8:31; 2 Pet. 2:4; Jude 6; Rev. 9:1-2, 11.

[98] Rev. 11:7; 12:9, 12; 13:1-4; 20:1-3.

[99] Job 38:17; 2 Pet. 2:4; Rev. 9:11; 17:8.

[100] 1 Pet. 3:19.

things creeping innumerable, both small and great beasts. There go the ships; there is that LEVIATHAN, WHOM THOU HAST MADE TO PLAY therein." Ps. 104:24-26

"Praise the Lord from the earth, YE DRAGONS AND ALL DEEPS." Ps. 148:7; Isa. 43:20

Although the Bible says God CUT Rahab, WOUNDED the Dragon and BROKE the heads of Leviathan, it also states that it is not yet time for all to be "punished". It says: "These wait all upon thee; that thou mayest give them their meat [karma] in due season. That thou givest them they gather," Ps. 104:26-28. This sounds like the negative ones are going to be given a second chance to choose the light over darkness. Perhaps some of the "dragons" will change and begin to "praise" God, however, according to the Biblical prophecies, others will war against the children of light, both inside and outside the Earth.[101] There will arise "deceivers, false prophets," and the "antichrist".[102] But, after the "weeds have been separated from the chaff", both the negative ones on the surface and the troublesome creatures from the Inner Earth, will be gathered up and chained.[103] Then the planet will finally experience a "millennium of peace"; a "Golden Age".[104]

Many people believe these Biblical descriptions are myths or fantasies, but perhaps each symbolic term depicts an actual being in the history of the early ages of Planet Earth. Other legends, texts, and scriptures, throughout China,

---

[101] Ezek. 26:19-21; Matt. 24:21-22; Rev. 8:7-11; 9:1-3, 11; 11:7; 12:12; 13:1-4; 16:13-14.

[102] Matt. 24:24; Rev. 12:9; 19:19.

[103] Matt. 3:12; Luke 3:17; Rev. 20:4.

[104] Rev. chapters 21 and 22.

India, Babylonia and Ancient America also describe technological wars, including underwater, as well as air battles, between the forces of light and dark. It was due to these atomic wars and their pollution that some of the different positive groups first moved underground. The Lemurians, Atlanteans, Mayans, Incas, the ten lost tribes of Israel, and the Essenes, are believed by many to have been taken inside the Earth, where they still reside. According to Dickhoff: "King Taupac Amaru escaped with forty thousand survivors from the Inca empire, after it had been taken over by the Spanish Conquistador, Pizzaro in 1533."[105]

The great mystics proclaim: "AS WITHIN, SO WITHOUT". Therefore, throughout this writing, I will assume that if there is an Inner Earth within the interior of the planet, there is also an inner core within each of us. And just as the Inner Earth is covered by the lakes, seas and oceans, so our inner core is covered by a sea of emotion. Manly P. Hall explains:

> "Man's lower nature was symbolized by a tremendous, awkward creature resembling a great sea SERPENT, or DRAGON, called LEVIATHAN. All symbols having serpentine form or motion signify the solar energy in one of its many forms. This great creature of the SEA therefore represents the solar life force imprisoned in water and also the divine energy coursing through the body of man, where, until transmuted, it manifests itself as a writhing, twisting monster--man's greeds, passions and lusts."[106]

---

[105] *Agharta*, Dickhoff, p 46.

[106] *The Secret Teachings of All Ages,* Hall, Philosophical Research Society, 3910 Los Fexiz Blvd. Los Angeles, CA 90027, 1928, p lxxxvi. (This description is also depicted in Job, chapter 41.)

Jesus instructed us to "resist not evil", Matt. 5:39, for all things will be corrected in due time. This is best described in the Biblical Job, chapter 41. I have reprinted the entire chapter because it so aptly describes mankind's pride and arrogancy, as well as the pride and arrogancy of the many beings who life within the Inner Earth. Again, "As within, so without".

> "Canst thou draw out Leviathan with a hook? Or his tongue with a cord which thou lettest down? Canst thou put an hook into his nose? Or bore his jaw through with a thorn? Will he make many supplications unto thee? Will he speak soft words unto thee? Will he make a COVENANT with thee? Wilt thou take him for a servant forever? Wilt thou play with him as with a bird? Or wilt thou bind him for thy maidens? Shall the companions make a banquet of him? Shall they part him among the merchants? Canst thou fill his skin with barbed irons? Or his head with fish spears? Lay thine hand upon him, remember the battle no more. Behold, the hope of him is in vain: shall no one be cast down even at the sight of him?
>
> None is so fierce that dare stir him up: who then is able to stand before me? Who hath prevented me, that I should repay him? Whatsoever is under the whole heaven is mine. I will not conceal his parts, nor his power, nor his comely proportion. Who can discover the face of his garment? Or who can come to him with his double bridle? Who can open the doors of his face? His teeth are terrible round about. His scales are

his pride, shut up together as with a close seal. One is so near to another, that no air can come between them. They are joined one to another, they stick together, that they cannot be sundered.

By his nessings a light doth shine and his eyes are like the eyelids of the morning. Out of his mouth go burning lamps, and sparks of fire leap out. Out of his nostrils goeth smoke, as out of a seething pot or cauldron. His breath kindleth coals, and a flame goeth out of his mouth. In his neck remaineth strength, and sorrow is turned into joy before him. The flakes of his flesh are joined together; they are firm in themselves; they cannot be moved. His heart is as firm as a stone; yea as hard as a piece of the nether millstone.

When he raiseth up himself, the mighty are afraid; by reason of breakings, they purify themselves. The sword of him that layeth at him cannot hold; the spear, the dart, nor the habergeon [coat of mail]. He esteemeth iron as straw, and brass as rotten wood. The arrow cannot make him flee; sling-stones are turned with him into stubble. Darts are counted as stubble; he laugheth at the shaking of a spear. Sharp stones are under him: he spreadeth sharp pointed things upon the mire. He maketh the deep to BOIL LIKE A POT; he maketh the sea a pot of ointment. He maketh a path to shine after him; one would think the deep to be HOARY. Upon earth there is not his like, who is made without fear. He beholdeth all high things; he is a king over all the children of pride."

The last few lines seem to indicate underwater explosions; perhaps a war in the Inner Earth. But more than a prophecy, this chapter in the book of Job is warning us of "pride and arrogancy", warning that "pride goeth before destruction", Prov. 8:13; 16:18. Blavatsky adds:

> "Although Christian theology has always connected both LEVIATHAN and BEHEMOTH with the devil, the expressions are metaphorical and have nought to do with the 'evil one'. . . . The LEVIATHAN is occult [hidden] science, on which one can lay his hand, but do no more, whose power and comely proportion God wishes to conceal."[107]

God allows the prideful beast, "Leviathan", to "play" in the oceans. What havoc he must still be reeking with the inhabitants of the Inner Earth. But will he eventually choose to serve the light or continue serving the forces of darkness. Every being is tested and all are given free will to choose who they will serve. The tests are not easy, for "Satan himself is transformed into an angel of light", 2 Cor. 11:14. In other words, he is a "shapeshifter". We need, therefore, to overcome our pride, arrogance and judgements and to learn discernment, 1 Cor. 2:14; Heb. 5:14. This is no easy task.

---

[107] *The Secret Doctrine,* Blavatsky, Vol. II, p 206; *Isis Unveiled*, Blavatsky, Vol. II, p 499.

CHAPTER 9
IS THERE A GOVERNMENT COVER-UP?

THE SECRECY CONCEPT:

It is becoming increasingly evident that many governments are aware of the polar openings, UFOs, and the extraterrestrials who operate them, as well as the innerterrestrials who live within the Inner Earth. Evidently this area lies beneath a huge sea of secrecy, the big "C" of conspiracy. But why do they want to keep such information a secret? Why were the "hole in the pole" photos confiscated? Why aren't the controversial land sightings being investigated? Why is the existence of UFOs and aliens, especially those seen diving into or emerging from the ocean, being denied? Why are such things as the face on Mars, which is not even on our planet, being denied? Why isn't the controversial moon landing being investigated? Is our government building a survival complex within the hollow moon or did they find the moon already inhabited by E.T.s, as claimed by Dr. Steven Greer and Howard Menger?[108] Is the government aware that all planets are hollow? Did the atomic submarine "Nautilus" really pass under the North Pole? Is this another coverup? Why did our government, namely the CIA and M15, put pressure on Robert Temple, author of *The Sirius Mystery*, for fifteen years after his book was published in 1976? What was the purpose of this "Sirius-secret" conspiracy? Why aren't we supposed to know that the Sirians, as well as the Pleiadians, and others, have visited our planet?

What is the advantage of such secrecy? Desiring to control mankind, they apparently want to use the knowledge of UFOs, Martians, Sirians, and the Inner Earth for their own

---

[108] *The High Bridge,* Howard Menger; *Disclosure*, Dr. Steven Greer.

purposes. Why? Perhaps it is due to the advanced technology involved, a technology which they could use to control the world. Perhaps it is because they want to take possession of this inner, paradisiacal land for their own survival meccas. Perhaps if there is a landmass near the North Pole, they are using it as a strategic-defense measure. Perhaps the don't want it known that "free energy" exists. Since the oil cartels have fought solar power and electric cars, how would they react to Inner Earth information about harnessing energy from the galaxy? Would such tactics topple the conspiracy's hold on us? How about the doctor and dentist cartels that suppress information on cancer and other cures as well as cavity prevention?

When I first began researching this subject, I was not aware of the degree to which it would disrupt my comfort zone and cause an inner sense of paranoia. Having taught spiritual classes for the past thirty years, during which I performed sacred ceremonies and learned to trust in God and flow through the difficulties of life, I was aware of the workings of the dark forces. I was naively unaware of the extent to which a huge governmental conspiracy was striving to control, manipulate, and brainwash mankind. Reading stacks of Internet material and various books, I was shocked to hear of such things as:

Aliens have invaded our planet and are ABDUCTING people and taking them underground to use for genetic experimentation

The underground facility, Montauk, ABDUCTS children for slavery and experimentation

The ETs who are doing the ABDUCTIONS have our government's approval

The Dulce (Mt. Archuleta) Underground Base network, below northwestern New Mexico, is a joint government and alien (reptilian) base

The Draconian Empire, reptilians from Alpha Draconis, have an underground base in New Mexico

The huge underground facility, Dreamland, centered below the Western front of the Rockies of Utah, stretches north to Idaho, east into Colorado, south to Arizona, and west into Nevada; it was Dreamland that originally drove out the Hopis

There is a reptilian base beneath Boynton Canyon in Sedona, Arizona

The Grays from Zeta Reticuli joined with the reptoids under Four Corners

The Ashtar-Bavarian Command, and such intelligent agencies as CIA and NSA, are aligned with the reptoidal Draconian and Gray Empire Collective

The Los Alamos nuclear-powered tunnel machine burrows by melting stone, leaving a smooth, glazed finish

Under the code name, "Noah's Ark," the New World Order is connecting the "Hidden Empire's" sub-city complexes

The "Unholy Six" (Epsilon Bootes, Altair Aquila, Capella, Zeta Reticuli, Rigel, and Bellatrix in Orion) of the Draco/Orion Empire are planning a take-over

Battles (Groom massacre, Dulce wars, and others) between our government and the Grays who turned against us

Human "men in black" chase, threaten, and kill people silence them quiet

I also found that different conflicting ET races have separate networks. Many of them are our positive helpers, while some are negative. The Sirians and Lyrans allegedly have bases below Egypt; the Centaurians, in the Maldive Islands; the Korendians, in Massachusetts; the Koldasians, in South Africa; The Zetas, in New Mexico and Nevada; the Venusians, in Brazil, Nevada and Tibet; and the Orions, in Iraq.

The newsletter, "The Hollow Earth Insider," states: "The New World Order plans to control entry into the ancient underground cities." Why? Because the Alternative 003 plans of these super-rich military-industrialist families, who first wanted to build survival bases on the moon, then on Mars, were sabotaged and taken over by the Draconians reptoids and the Grays. Their new plan, "Noah's Ark Project", is to build underground cities and colonies within the Earth.

The government is even restricting caves. David Crenshaw, in his newsletter, "The Hollow Earth Insider Research Report" states:

> "Unknown by most people as of November, 18, 1988, both the Senate and House of Representatives passed Public Law 100-691, commonly referred to as the Federal Cave Resources Protection Act of 1988. . . . The government has decided that caves located on 'Federal Lands' are a natural resource and must be protected. . . . Another interesting provision, under Section 5, 'Confidentiality of Information Concerning Nature and Location of Significant Caves' states: 'Information concerning the specific location of any significant cave may not be mae available to the public under section 552 of title, United States Code."[109]

Whenever anything suspicious is found in a cave, the authorities have it dynamited to close off the entrance. Why this severe sabotage? What are they trying to hide? Perhaps we need to demand that technology be made available to the

[109] *The Hollow Earth Insider,* volume 3, #5. onelight.com

public, so we can protect ourselves against the government making us into slaves of the system.

Erich Von Daniken, in his book *Gold of the Gods*, shows the copy of a deed with the notation: "With the issue of this notarial title-deed dated on 21 July, 1969, the caves beneath Ecuador became the property of Juan Moricz. Moricz put them under state control, which will smooth the way for future research.[110] Great idea! Sounds like Ecuador will be a New-Age mecca.

With this depressing negativity, fortunately, I also found some good news:

There are one hundred twenty cities in the spiritual Agharta Network

According to the *Ramayana*, the great avatar Rama, was "an emissary from Agharta"

The *Egyptian Book of the Dead* describes the "tunnels of Ra"

The Emerald Tablets of Thoth describes tunnels in the "Halls of Amenti" under the Great Pyramid

Osiris is a subterranean god who rules the underworld

The Telosians are allied to at least 100 human-occupied underground cities throughout the world, most of which are in conflict with the reptilian 'Draco' societies and their Gray collaborators

The people of Telos are busy planning survival centers for refugees. One is near Prescott, another near Jenny Lake, Wyoming, close to the Tetons

The spiritual races, the Uighers, Naga-Mayas, and Quetzals, inhabit Telos

There is a major (positive) Federation base in California

---

[110] *Gold of the Gods*, Erich Von Daniken, G. P. Putnam's Sons, NY, 1972, p 3.

The Mu-Suvians, or Hav-musuvs, of the Mojave and Death Valley region, are in direct opposition to the reptilian species operating from the Dulce Network

The Andromeda/Pleiadian Empire are non interventionalists who have always monitored and protected us

American Intelligence agencies such as COM-12 are allied with the benevolent ones of the Federation

According to Sharula, of Telos, located under Mt. Shasta in California, the Telosians are at odds with the Grays. She was told by Adama, the spiritual leader of the Melchizedek Lodge, that the Grays are being asked to leave the planet, by force, if necessary. The last Gray stronghold that is to be taken care of is the one near Los Alamos, New Mexico, which is the eastern segment of the Dulce Hub.

According to Branton, the government's plan backfired. He says:

> "In order to remain hidden from the public the Grays made a treaty with the United States Government to buy time for them to take control of the minds of the human agents. They allowed the CIA to possess a few of the upper levels and expand them, but the base workers literally lost their minds to the Alien controllers via implantation."[111]

The "Blues", also called the "Children of the Feather", were one of several human-like races that warned our government about having anything to do with the Grays. The present-day Hopis, both surface and subsurface dwellers,

---

[111] *Dreamland in the Rockies,* Branton, p 6; MOVAVE/COSCON; http://www.think-aboutit.com/branson/mojave.htm

still consider the Grays of the Four-Corners systems and elsewhere to be their enemies. Hank Krastman says the residents of Palatkwapi, under Arizona, maintain some interaction with the new-Mayan races of Mt. Shasta who maintain a large underground city near Prescott, Arizona. They have also interbred with members of the Pleiadian Federation.

Other information on the bright side is given by Branton:

> "If the alternative 003 forces could be deprogrammed and their loyalties turned toward the Federation and against the Draconians, they can be just as much a force for good as they have been for evil. . . It is possible to use the telepathic implants by which an abductee's mind is influenced or 'controlled' by the reptilian collective, and turn it back in their face, using it as a weapon against them. The reptilians are taking a chance by connecting humans into their collective via the implants, and I have found that it is possible to turn things around and use the implants to probe their darkest secrets, intimidate them, possibly communicate with the hybrids, and virtually raise all kinds of hell with the reptilian collective mind -- providing that all this is done under the direction of Divine Grace."[112]

So, perhaps there will be a pleasant ending to this

---

[112] *Dreamland in the Rockies,* Branton, p 7; MOVAVE/COSCON; http://www.think-aboutit.com/branson/mojave.htm

battle between the forces of light and darkness. But, like all things on Earth, it seems there is the proverbial "darkness before the dawn." We are currently in a period of "tribulation," but a "millennium of peace" -- a "Golden Age" -- is right around the corner. It has been explained to me by my mentor, the Ascended Master[113] Saint Germain, that the "dragon" has already been slain on the inner planes, but, like all beings of the lizard family, it still has the propensity to flail its tail around to do as much damage as possible before it take its last breath. I believe its breath is already becoming shallow. Napoleon, Hitler and "Babylon" [Saddam Hussein] have already fallen and "divine intervention" is standing by. (Nostradamas predicted three powerful "anti-Christs" - Napoleon, Hitler, and Sadam Hussein. He called Napoleon "the man with an unknown name"; he called Hitler, "Hister", and called Sadam, "Madus". Scrambling letters was a common practice in that day.)

The Bible promises:

> "For then shall be great tribulation, such as was not since the beginning of the world to this time, no, nor ever shall be. And except those days should be SHORTENED, there should no flesh be saved: but for the elect's sake, those days shall be SHORTENED. . . Immediately after the tribulation of those days shall the sun be darkened, and the moon shall not give her light, and the stars shall fall from heaven, and the powers of the heavens shall be shaken. And then shall appear the sign of the Son of Man in heaven: and then shall all

[113] ASCENDED MASTER is a term applied to those who have "graduated" from their earthly cycle of rebirth and ascended into a higher plane of existence. Some of the Masters then serve at cosmic levels, while others continue to serve this planet. Saint Germain continues to serve here.

> the tribes of the earth mourn, and they shall see the Son of Man coming in the clouds of heaven with power and great glory." Matt. 24:21-22, 29-30

This will be further explained in chapter 11. We will soon enter into a bright new "Millennium of Peace." Perhaps then we will finally achieve a government "uncover-up".

CHAPTER 10
INNER EARTH INFO FROM ENLIGHTENED WRITERS

THE SPIRITUAL CONCEPT:

1. THE EMERALD TABLETS OF THOTH

The Emerald Tablets of Thoth are believed to have been compiled around 20,000 B.C. They explain how the Atlantean priest-king, Thoth, after the sinking of his mother country, Atlantis, founded a colony in ancient Egypt. The Atlanteans also spread in many other directions to teach and uplift mankind, but Egypt was the headquarters of Thoth.[114]

Why did Atlantis sink? Thoth says the wisdom of the "children of light", from whom the Atlantean "children of men" had learned, was gained from the star-born races,[115] but at a later time, they were infiltrated "by the spawn of a lower star".[116] Thoth says that due to this infiltration, the "children of light":

> "Then, for a dwelling place, FAR 'NEATH THE EARTH CRUST, BLASTED GREAT SPACES they by their power, spaces apart from the children of men. . . . Side by side then, placed they OTHER SPACES, filled them with Life and with Light from above."[117]

---

[114] *The Emerald Tablets of Thoth*, translated by Doreal, Brotherhood of the White Temple, 7830 Oak Way, Sedalia, CO 80135, 1939, Tablet I, p 4.

[115] The *Emerald Tablets of Thoth*, Tablet XI, p 42.

[116] *The Emerald Tablets of Thoth*, Tablet I, p 3.

[117] *The Emerald Tablets of Thoth,* Tablet II, p 6; Tablet V, p 20; Tablet VI, p 22.

They built what they called the "Halls of Amenti".[118] In these "Halls", Thoth and the "children of light" periodically--approximately one hundred years out of each thousand--lay beneath the "Flower of Life", where the life force of this planet is concentrated. In this manner they renewed their bodies and retained their immortality.[119]

Thoth also spoke of the dark beings, saying:

> "In the form of man moved they amongst us, but only to sight were they as are men. SERPENT-HEADED when the glamour was lifted . . . crept they into the Councils . . . Beware, the serpent still liveth."[120]

Upon his arrival in Egypt, Thoth built the Great Pyramid of Giza to conceal the subterranean entrance into the "Halls of Amenti", the underworld where "the great king sits upon his throne".[121] After placing ancient records and technological instruments of Atlantis in the Great Pyramid, Thoth erected the Sphinx over his huge buried spacecraft. He said it would be needed when invaders from outer space attacked the Earth.[122] Thoth gave detailed instructions of how to find these buried treasures and even described the layout of the interior of the Great Pyramid, and the Inner-Earth chambers beneath it.[123] He related:

---

[118] *The Emerald Tablets of Thoth*, Tablet V, p 20.

[119] *The Emerald Tablets of Thoth*, Tablet I, p 1; Tablet II, p 7.

[120] *The Emerald Tablets of Thoth*, Tablet VIII, p 32.

[121] *The Emerald Tablets of Thoth*. Tablet I, p 2.

[122] *The Emerald Tablets of Thoth*, Tablet V, p 21.

[123] *The Emerald Tablets of Thoth*, Tablet V, p 21; Tablet XI, pp 42-43.

> "Raised I high over the entrance, a doorway, a gateway leading down to Amenti. . . . Raised over the passage, I, a mighty pyramid, using the power that overcomes Earth force. Deep and yet deeper placed I a FORCE-HOUSE or CHAMBER."[124]

The apex of the pyramid drew in force from the ethers. Thoth says the keys to Amenti are hidden within the pyramid. When Egypt was chosen to be the headquarters of the new race-wisdom, the natives were trained by the survivors of Atlantis, who acted as "custodians and watchers".[125] Thoth appointed pyramid "guards" to watch over the treasures buried in and around the Great Pyramid and the Sphinx. The descendants of these "guards" later became "pyramid priests," while Thoth was deified as the "God of Wisdom". The "Halls of Amenti" became known as the "underworld, the halls of Osiris and other gods, where the soul passes for judgment after death".[126] This, however, was only one portion of the "Halls of Amenti", other portions of it served as home to the "children of Light".

During later ages, Thoth incarnated three times. In his last incarnation he was known as Hermes, Trismigistis. (Trismisgistis means thrice born.)

Concerning the actual "emerald tablets" that were left in the pyramid by Thoth, the preface of *The Emerald Tablets of Thoth* states:

> "They consist of twelve tablets of

---

[124] *The Emerald Tablets of Thoth*, Tablet I, p 4; Tablet 2.

[125] *The Emerald Tablets of Thoth*, Tablet II, p 7.

[126] *The Emerald Tablets of Thoth,* preface, p 1; Tablet I, p 1-3.

> emerald green, formed from a substance created through alchemical transmutation. They are imperishable, resistant to all elements and substances. . . . Upon them are engraved characters in the ancient Atlantean."[127]

The tablets remained in the Great Pyramid for many thousands of years. Then around 1300 BC, when Egypt was disintegrating, priests were sent to other parts of the world. Thoth says the "pyramid priests" carried with them the "emerald tablets" as a "talisman by which they could exercise authority over the less advanced priest-craft of races descended from other Atlantean colonies." This was a measure to retain the purity of the teachings. The authorized priests migrated to South America to share the tablets with the Mayans. In the tenth century, the Mayas, who settled in Yucatan, brought the tablets with them and placed them beneath one of the altars. After the Spanish conquest, the Mayan cities were abandoned, but the "emerald tablets" remained hidden under the altar.[128]

The man who found and translated them was Maurice Doreal, a member of the Great White Lodge, an organization that still works through the "pyramid priesthood". He was instructed to recover and return the tablets to the Great Pyramid. He did so in 1925, but before returning them, he was given permission to translate and retain a copy of the wisdom engraved thereon. The translation, divided into fifteen sections, was held in secret until 1939, at which time the first thirteen tablets were released to the public by the "Brotherhood of the White Temple." Numbers fourteen and fifteen are so powerful that they were held, then released at a later time. These fifteen

---

[127] *The Emerald Tablets of Thoth,* preface, p 2.

[128] *The Emerald Tablets of Thoth*, preface, p 1.

tablets can be purchased from the "Brotherhood", which is still an active organization.[129]

Thoth gives an interesting prophecy for the future, promising divine intervention:

> "Ages to come shall see revival of wisdom to those who shall inherit thy place on this star. . . . Many who are bound in darkness shall strive to hold others from Light. Then shall there come unto man the great warfare that shall make the Earth tremble.[130]

Thoth also speaks of Shamballa and gives us the key to enter the Inner Earth:

> "Now give I the Key to Shamballa, the place where my Brothers live in the darkness: darkness but filled with Light of the Sun -- darkness of Earth, but Light of the Spirit, guides for ye when my day is done."[131]

The remainder of these instructions is contained in the sacred Tablet XV. Due to the power of its secrets, I have not gained permission to share it in this writing. It can be, however, be ordered from the "Brotherhood of the White Temple".

De Lucia, in *Hollow Earth in the Puranas* relays a message that substantiates the underground chambers under the Great Pyramid of Giza:

---

[129] Brotherhood of the White Temple, 7830 Oak Way, Sedalia, CO 80135.

[130] *"The Emerald Tablets of Thoth*, Tablet XII, p 48.

[131] *The Emerald Tablets of Thoth*, Tablet XV, p 55.

"Since the declassification of the new ground-penetrating radar two years ago, the most staggering data has emerged of complex and labryrinthine underground systems in various parts of the world. In similar fashion, the SIRA radar was deployed in Egypt as early as 1978, mapping an extraordinary subterranean complex beneath the Egyptian pyramids. Arrangements made with President Sadat of Egypt, resulted in three decades of top secret excavations to penetrate the system. At a recent meeting in Australia, one of the key scientists on the Giza project, Dr. Jim Hurtak[132]showed film footage of work in progress called, CHAMBERS OF THE DEEP, due to be released at the end of the century [2000]. The film reveals the discovery of a vast megalithic metropolis, 15,000 years old, reaching several levels below the Giza plateau. While the rest of the New-Age speculates about a hidden chamber under the left paw of the Sphinx, the legendary 'City of the Gods', lays sprawled beneath. Complete with hydraulic underground waterways, the film shows massive chambers, the proportions of our largest cathedrals, with enormous statues, the size of the Valley of the Nile, carved in-situ. Researchers, risking their lives with lights and cameras, carefully negotiated rubber dinghies across subterranean rivers

---

[132] Dr. J. J. Hurtak founded The Academy for Future Science, and wrote *The Book of Knowledge: The Keys of Enoch* in 1973. Current address is: P. O. Box FE, Los Gatos,CA 95031.

and kilometer-wide lakes, to penetrate sealed chambers beyond. Already, remarkable caches of records and artifacts have been found. It is the legacy of a civilization and a technology way beyond our own. A technology capable of creating a vast underground city, of which the sphinx and pyramids are merely the surface markers." (Paul White)[133]

2. H. W. PERCIVAL

While researching the subject of the Inner Earth, I read a marvelous book, *Thinking and Destiny*, by H. W. Percival, who was born in 1868. At age 24, he attended one of Madam Blavatsky's Theosophy meetings. He considered becoming a pupil of one of the "Mahatmas" (enlightened men or Ascended Masters), but decided to walk the spiritual path without a teacher. The following year, he was totally enlightened. He explained his experience as:

> "Light, greater than that of myriads of suns opened in the center of my head. In that instant or point, eternities were apprehended. There was no time. Distance and dimensions were not in evidence. Nature was composed of units. I was conscious of the units of nature and of units as Intelligences. Within and beyond, so to say, there were greater and lesser Lights; the greater pervading the lesser."[134]

---

[133] *Hollow Earth in the Puranas* De Lucia, pp 45-46.

[134] *Thinking and Destiny,* H. W. Percival, The Word Foundation, Inc., NY, 1946. Current address: P. O. Box 17510 , Rochester, NY 14617.

Throughout his enlightening book, he refers to the Inner Earth:

> "Then came a Third Civilization . . . there came peoples from inside the crust of the earth. Some were descendants of people who had sought refuge there from the corruption under the rule of the handworkers, had escaped the cataclysm on the outer crust and had increased in numbers. Others were those who had fled from an Inner Earth toward the outer crust. They were the descendants of those who had failed, who had lost their perfect bodies and had taken the path of death and re-existence. . . . However, some of the better kind of doers had been led to safety chambers in the interior of the earth crust, where they propagated and continued to live through those ages. They came out, subdued the savages and taught them husbandry, the working of woods, metals and stones and weaving of grasses. , , , They traveled a great deal on and under water in boats which they propelled by motive power obtained through the water. They knew how to harden water without freezing, and used the transparent mass to fill apertures and to admit light. They extracted while under water all the air they needed for breathing. They had access to subterranean waterways and to vast oceans within the earth crust. . . . The people were not all of one color; some were white, some red, some yellow, some green, some blue or violet; and they were of light and dark shades and combinations of these colors. . .

. Then began a Fourth Civilization . . . Some of the decadent inhabitants of the former earth survived the submersion and wandered, sailed or drifted to mountains whose tops were above the water. New exiles came out of the chambers of the earth crust. . . . There appeared then among these struggling groups, men of a kind superior to them. They came from the interior of the earth and were of superior intelligence, so that the savages saw that it was useless to fight against them. These men taught the savages the use of fire and how to make rude implements, and established primitive social order. They gave some grains to the rude people, showed them how to grow them and taught them to build houses. Gradually, they taught the people to domesticate some animals, to weave, to work in metal, and to build with stones. . . . They had underground passages through which they moved to distant parts of the earth within a day; for in these passages they created a current that went with them so that they did not meet the friction of the air."[135]

Percival further explained:

"During the immense periods which have passed in this Fourth Civilization, there have been many changes in the structure of the earth's crust, in the surface distribution of land and water, in the inclination of the poles, in the climates in

[135] *Thinking and Destiny*, pp 323-328.

different parts, in the magnetic and electric currents in the earth, the water, the air, and the starlight. . . . From the standpoint of man with his conception of length, breadth and thickness as three dimensions, the earth, if it could be seen, would appear as a sponge-like spherical crust, between three outer spherical layers and three inner spherical layers. The earth crust has an outer and an inner surface. The distance between these varies from about two hundred to eight hundred miles. . . . Between the outer and the inner skin of this solid layer are large and small subterranean chambers in which there are fires, airs, oceans, lakes and rivers, which often differ in their qualities and colors from the things seen on the outer crust. Minerals, plants and animals are different in their forms and habits from those known to human beings. Forces are operative there which are dormant on the outer crust. Gravity and other forces are changed from what they are as known to man. Many peoples in various degrees of development exist between the outer and the inner skin of this layer. They differ as to color, size, features and weight. To some of them the crust is a transparent as the air is to men. To others it is opaque, but they have a diffused earthlight by which they see. To these many races, in the different strata of the crust, their environment is a world. The beings inside and beyond the crust are of the nature-side or of the intelligent-side. Each set has its own world in which the constituents, requirements and possibilities are different. Beyond the outer and the inner skin, conditions are such as to be incomprehensible at present."[136]

3. GUY W. BALLARD:

Guy W. Ballard was also totally enlightened and had studied with the Ascended Master Saint Germain on the inner planes for thirty-three years. He compiled the "green books", *The Saint Germain Series*, under the pen name Godfre Ray King. The book gave some very interesting details concerning his experiences in the subterranean temples and cities. The ten "green books" encourage the reader to share them and to spread these messages world-wide.[137]

Guy Ballard related that in the fall of 1930, the Ascended Master Saint Germain came to him on many occasions, each time offering him a drink of spiritual elixir, then taking him etherically inside specific temples that were either within a mountain or underground.

When entering the "Retreat" within the Grand Teton Mountain, Ballard said Saint Germain merely touched a huge boulder at the top of the mountain and it opened as a doorway to reveal a massive bronze door and a solid rock stairway leading downward. After descending some two hundred feet, they came to a circular area and stepped into a tube elevator that lead them two thousand feet into the heart of the mountain. Saint Germain said the temple had been there for twelve thousand years, since the sinking of

---

[136] *Thinking and Destiny,* pp 518-519.

[137] *The "I Am" Discourses,* Vol. 3 of *The Saint Germain Series*, Saint Germain Press, IL, 1940, pp 245, 325, 335; *Ascended Master Light* Vol. 7, Saint Germain Press, IL, 1938, pp 127-128; *The "I Am" Discourses*, Vol. 8, Saint Germain Press, IL, 1942, p 127; *The "I Am Discourses*, Vol. 9, Saint Germain Press, IL, 1949, p xi, 365.

Atlantis.[138]

On another occasion, Saint Germain took Mr. Ballard to a subterranean temple in the ancient city of Mitla in the state of Oaxaca, Mexico. Here, Saint Germain removed a pile of rocks to reveal a rose-colored granite cube. He touched it and, as it turned, it revealed an opening with steps leading downward.[139]

When Saint Germain brought Mr. Ballard to a buried city in the Amazon, he took him through the water. Ballard was surprised that he could breathe in that element. According to Saint Germain, the cities in that area had been built between fourteen and twelve thousand years ago.[140]

Saint Germain explained:

> "Beloved Ones! does it sound incredible to you, when I say the Masters of Light and Wisdom have passages through the Earth in all directions, just the same as you have highways for your automobiles in going from coast to coast on the Earth? Did you understand the atomic structure of the Earth, you would not feel this such an incredible thing; for those Great Ones who have assisted the progress of humanity from the beginning, have but to use certain Rays and they walk through the Earth as easily as you would walk through the water; different however, in that they leave the opening behind them while in your walking through

---

[138] *Unveiled Mysteries*, Vol. I, Godfre Ray King, Saint Germain Press, Schaumburg, IL, c 1939, pp 73-76.

[139] *Unveiled Mysteries*, Vol. I, p 130.

[140] *Unveiled Mysteries,* Godfre Ray King, p 170-172.

> the water, it closes behind you and the path is not apparent."[141]

While visiting in Denver, Colorado, Mr. Ballard and his friends were taken by Saint Germain into a special cave, called the "Cave of Symbols". There, they entered an elevator that took them down a hundred feet to a magnificent temple.[142] They were shown a perfected radio through which one could hear broadcasts from other planets of our Solar System, from the Etheric Belt around the Earth, from anywhere on Earth, and from the interior of the Earth. While there, they tuned the "super" radio into the interior of the Earth. A being in that area explained their temperate climate, their soft white light, and the fact that their atmosphere retains an equal pressure.[143] Saint Germain proclaimed:

> "The idea that the center of the Earth is a mass of fire is entirely erroneous. Within the crust of the Earth for a certain depth, there are conditions of the Fire-element, acting; but within the center of the Earth itself, there are Self-conscious Individual Beings, who through many cycles of Work and Self-effort have mastered control of certain forces. . . .Why should it not be a perfectly natural, normal condition that would permit some of their number to be the Instructors of those, who are working with forces at the center, as well as at the circumference of the

---

[141] *The "I AM" Discourses,* Vol. 3, Godfre Ray King, Saint Germain Press, Schaumburg, IL, c 1940, Saint Germain Foundation 2000 printing, p 202.

[142] *The Magic Presence,* Vol. II, Godfre Ray King, Saint Germain Press, Schaumburg, IL, c 1935, pp 59-69, 140.

[143] *The Magic Presence*, Vol. II, Godfre Ray King, pp 77-80.

planets?"[144]

Mr. Ballard was also shown other subterranean cities that had been hermetically sealed. One was 125 feet from beneath the surface, while another lay on the bed of the Atlantic ocean.[145] These teachings, within the set of "green books", are the most interesting and spiritually uplifting of any I have encountered.

4. A. D. K. LUK:

This concept of beings working there who have control over the forces is detailed by A.D.K. Luk, who served as secretary for the Ballards during the 1930's and 1940's. She later published a series of "yellow books." In them, she states:

> "When the planet Earth was to be created, Virgo and Pelleur[146] (Twin Rays) offered their services to produce or supply the substance of the earth element. . . . This earth substance was created from the atoms formed by drawing (pure) electrons around a central core . . . and the colors of the seven rays played through it producing an iridescent radiance. . . . In the beginning, the Earth's substance was crystalline, pure and iridescent, and the glowing radiation

---

[144] *The Magic Presence*, Vol. II, Godfre Ray King, Saint Germain Press, Schaumburg, IL, c 1935, Saint Germain Foundation, 2002 printing, pp 81-82.

[145] *The Magic Presence*, p 62, 326-327.

[146] Pelleur is a Cosmic Being who is the directing intelligence of the activities in the center of the Earth. Virgo is the Goddess of Earth who provided the substance of earth and the elements of the physical body. Her substance also holds the water element. (For further details, see *Law of Life, Book II, pp 338-339.)*

from the very center of the Earth was easily discernable, pouring out through the Earth and its atmosphere, allowing the bodies of the people to absorb the radiant substance and to live, move and have their being within that atmosphere of purity and peace. This gave the people assistance in the speedy fulfillment of their own divine plan. As one looks down through clear water and is able to observe the submarine life, so could any member of the race then embodies look into the depths beneath the Earth's surface and see the Salamanders disporting themselves in the pink and golden rays from the center of the Earth. . . . The Cosmic Being Pelleur's Focus is in the center of the Earth, where there is the Sun of 'Even Pressure' and continuous soft light; with vegetation of ivory and pink. . . . Pelleur's radiance from the center permeated the entire Earth. . . . The center of the Earth is a place of great perfection and a great focus of light. There is no night there but a soft white radiance of even pressure and perfect climatic conditions all the time. It is an harmonious and very peaceful place. . . . In some past age some people living on the surface of the Earth were driven by the discord of the surface into the interior because they refused to slaughter the invading hordes. These people were determined to put the authority of the light first. The Cosmic Law opened up the Earth so they could go into the center and live in Pelleur's domain. There is a delightful semi-tropical climate, no seasons of heat and cold not extreme temperatures. These Beings live there in harmony, there is no discord or decay. . . . When lifestreams take embodiment there that do have some discordant karma, they are consulted about it and with their permission it is consumed. . . . From Pelleur's region much assistance is being given on the surface of the Earth. Light rays are being directed from there to the surface and in cooperation with the Cosmic Light they are assisting in the cleansing activity of the Earth and the bringing forth of perfection. His activity is a balancing action. . . . Beloved Pelleur also had to lower the vibrations of the gravity pill of Earth and create a centripetal force within the Earth which would correspond to the denser type of creation which mankind had drawn about it on Earth's surface. . . . It was once said that there were subterranean tunnels connecting the Grand Teton Retreat, the Cave of Symbols and Mount Whitney."[147]

5. MANLY P. HALL:

In his book, *Man, Grand Symbol of the Mysteries,* Hall states:

"The most secret rituals of ancient initiation were performed in subterranean crypts, which were called 'Caverns of the Mysteries' [shamans also go into the Inner Earth for their ceremonies]. . . . The cave-dwelling Rishis of

---

[147] *Law of Life, Book II,* A. D. K. Luk Publications, OK, 1972, pp 207, 339, 451; *Law of Life and Teachings by Divine Beings, Book I*, A. D. K. Luk, A. D. K. Luk Publications, OK, 1978, p 144; *Law of Life and Teachings by Divine Beings, Book* 2, 426-427.

India are the seven original beings, who in the mythology of Central America, were the cave-born progenitors of the races."[148]

Manly P. Hall's book, *The Secret Teachings of All Ages*, explains that according to Kabalistic legend, CHiram was a descendant of Samael, the Spirit of Fire. Among his ancestors was the immortal Tubal-Cain, the forger of fire, who first tempered metal. CHiram built the "Molton Sea," a magnificent laver of purification, supported on the backs of twelve oxen. Hall says:

> "This huge basin, thirty cubits in circumference, was to be made of a single casting. But three craftsmen, disgruntled because CHiram [Hiram Abiff] refused to promote them, secretly poured water into the mold. When the taps of the furnace were opened and the flaming metal ran into the mold, a great explosion took place. Clouds of steam arose and molten metal, hurled in every direction, spread death and confusion. Though broken-hearted at the failure of his great ambition, CHiram stood courageously in the path of the streaming mass. Suddenly, a voice from the midst of the stream and smoke cried out: 'I have rendered thee incorruptible. Cast thyself into the flames.' CHiram thereupon plunged through the Molten Sea into THE CENTER OF THE EARTH where dwelt his fiery ancestor, the first worker in metals."[149]

---

[148] *Man, Grand Symbol of the Mysteries,* Manly P. Hall, The Philosophical Research Society, CA, 1972, p 160.

[149] *The Secret Teachings of All Ages*, Hall, opposite page CLIII.

Hall explains the esoteric symbolism:

> "The soul, constructed from an invisible fiery substance, a flaming golden metal, is cast by the Master Workman, CHiram Abiff, into the mold of clay (the physical body) and is called the Molten Sea. The temple of the human soul is built by three Master Masons personifying Wisdom, Love, and Service, and when constructed according to the Law of Life the Spirit of God dwells in the Holy Place thereof. The Soular Temple is the true Everlasting House, and he who can raise or cast it is a Master Mason indeed."[150]

6. ZECHARIA SITCHIN:

Sitchin, in his book, Stairway to Heaven, tells of Alexander, the Great's visit to the Inner Earth to seek immortality:

> "Alexander went to the Great Oasis, seat of the oracle of Ammon There, the god himself (so legend says) confirmed Alexander's true parentage. Thus affirmed, the Egyptian priests deified him as Pharaoh, thereby, his desire to escape a mortal's fate became not a privilege, but a right. . . . After a pleasant stay, the queen (Hatshepsut) agreed to reveal to Alexander, as a parting gift, the secret of 'the wonderful cave where the gods congregate.' Following her directions, Alexander found the sacred

---

[150] *The Secret Teachings of All Ages,* Hall, pp CLXXV-CLXXVI.

> place. He entered with a few soldiers, and saw a starlit haze. And the rooftops were shining, as if lit by stars. The external forms of the gods were physically manifest, a crowd was serving them in silence. . . . The sight of the 'reclining figures', with eyes emitting beams of light, made Alexander stop short. Were they too gods, or deified mortals? He was then startled by a voice; one of the 'figures' had spoken up. . . . Alexander was invited in to see the Creator and Overseer of the entire universe.' He went within and saw a fire-bright haze and, seated on a throne, the god whom he had once seen worshipped by men in Rokotide, the Lord Serapis. (In the Greek version, it was the god Dionysus.)"[151]

Alexander asked the gods for immortality, but was told that although he would die young, his name and deeds would be immortal. Very disappointed, Alexander left the cave and sought other sages. According to one version, he visited the biblical Enoch, who resided atop a holy mountain. Alexander saw a stairway with 2500 golden slabs leading to a cavern; "Paradise, the Land of the Living, the abode of the saints." After going there, Enoch told Alexander: "Do not pry into the mysteries of God." Again Alexander left in great disappointment.[152]

Sitchin further explains the Egyptian writings about the underground "Paradise of Seker":

> "To reach it in the House of Fire,

---

[151] *The Stairway to Heaven* Zecharia Sitchin, Avon, NY, 1980, pp 20-22.

[152] *The Stairway to Heaven,* pp 22-23.

the Great House of Two, he must enter the Amen-Ta, the Hidden Land of Seker, God of the Wilderness. It is a domain described as a fortified circle. It is the SUBTERRANEAN LAND OF DARKNESS, reachable by entering into a mountain and going gown spiraling hidden paths protected by secret doors. It is the fourth division of the Duat which the king must now enter, but the mountain entrance is protected by two walls and the passage between them is swept by flames and manned by guardian gods. When Ra himself had arrived at this entrance to the Hidden Place, 'he performed the designs'--followed the procedures--'of the gods who are therein by means of his voice, without seeing them.' But can the king's voice alone achieve for him admission? The texts remind the challenger that only 'he who knoweth the plan of the hidden shaftways which are in the Land of Seker,' shall have the ability to journey through the Place of Underground Passages and eat the bread of the gods. . . . 'The MOUTH OF THE EARTH opens for thee, the eastern door of heaven is open for thee,' the gods of the Duat announce to the king. He is reassured that though he enters the mouth of the earth, it is indeed the Gateway to Heaven, the coveted eastern door. The journey in the fourth and following hours leads the king through CAVERNS AND TUNNELS where gods of diverse functions are sometimes seen, sometimes only heard. There are UNDERGROUND CANALS, on which gods move about in silent barques.

> There are eerie lights, phosphorous waters, torches that light the way. Mystified and terrified, the king moves on, toward 'the pillars that reach the Heaven.'"[153]

7. JANE ROBERTS:

Jane Roberts is the channel for a being named Seth, who speaks of Inner Earth beings which he calls "Lumanians". In her book *Seth Speaks*, Seth relates:

> "Their vast technology and their great civilization was largely UNDERGROUND. . . . CAVES were not just places of protection in which unskilled natives squatted. They were often DOORWAYS to and from the cities of the Lumanians. . . . mechanically created channels that reached behind them."217,

> "While the civilization of the Lumanians was highly concentrated, in that they made no attempt to conquer others or to spread out to any great extent is true, they did set out, over the centuries, outposts from which they could emerge and keep track of the other native peoples. These outposts were constructed underground. From the original cities and large settlements there were, of course, UNDERGROUND connections, A SYSTEM OF TUNNELS, HIGHLY INTRICATE AND BEAUTIFULLY ENGINEERED. Since these were an aesthetic people, the walls were lined with

[153] *The Stairway to Heaven*, pp 52-53.

paintings and drawings, and sculpture was also displayed along these inner byways. There were various escalated systems, some conveying people of foot, some conveying goods. It was not practical to construct such tunnels to the many outposts, however, which were fairly small communities and relative self-supporting; some were a good distance away from the main areas of commerce and activity. These outposts were situated in many scattered areas, but there were a fairly large number of them in what is not Spain and the Pyrenees. There were several reasons for this, one having to do with the existence of rather giant-sized men in the mountain areas. Because of the timid nature of these [Lumanian] people, they did not enjoy outpost existence, and only the bravest and most confident of them were given such an assignment, which was temporary to begin with. . . . These caves, again, served as doorways opening outward, and often what seemed to be the BACK OF A CAVE was instead CONSTRUCTED OF A MATERIAL OPAQUE FROM THE OUTSIDE BUT TRANSPARENT FROM THE INSIDE. . . . There were frequent PEEPHOLES, so to speak, through to the surface, from which they could make observations, and cameras situated there that kept the most precise pictures not only of the earth, but of the stars. Of course, they had complete records of underground gas areas and intimate knowledge of their inner crusts, keeping careful watch upon and anticipating earth tremors and faults. . . . Some of the caves, particularly in certain areas of Spain and the Pyrenees, and some earlier ones in Africa, were artificial constructions. Now these people MOVED MASS WITH SOUND. . . This is how their tunnels were originally formed."[154]

Roberts adds a note under this channeling, stating:

"Seth gives no dates for the Lumanian civilization. It's interesting to note, however that in late July, 1971, about eight months after this session, newspapers carried the story--with photographs--of the unearthing of a 'massive' subhuman skull in a cave in the French Pyrenees Mountains, very close to the Spanish border. (The skull is at least two hundred thousand years old, and represents a race not identified before. It is now tentatively thought that several primitive races existed in Europe at that time. The period predates Neanderthal Man, and marks the start of the next-to-last Ice Age. The region in southern France is noted for its many caverns, easily eroded out of the limestone bedrock by flowing water."[155]

8. JOHN MILTON:

John Milton in *Paradise Lost and Other Poems*, writes about Satan's eviction from heaven, his descent into the BOTTOMLESS PIT at the CENTER OF THE EARTH, and of his building a BRIDGE to overcome the GULF, so he

---

[154] *Seth Speaks*, Jane Roberts, New World Library, Amber-Allen Publishing, CA, 1972, pp 214-224.

[155] *Seth Speaks*, pp 218.

could come up the surface of the earth to influence its residents. In brilliant poetic form, which has been a classic since 1640, Milton says:

"And crystal wall of heaven, which opening wide,
Rolled inward, and a spacious gap disclosed
Into the wasteful DEEP; the monstrous sight
Struck them with horror backward, but far worse
Urged them behind; headlong themselves they threw
Down from the verge of heaven, eternal wrath
Burnt after them to the BOTTOMLESS PIT.
Hell heard the unsufferable noise, hell saw
Heaven running from heaven and would have fled
Afrighted; but strict fate had cast too deep
Her DARK FOUNDATIONS, and too fast had bound.
Nine days they fell; confounded CHAOS roared,
And felt tenfold confusion in their fall
Through his wild anarchy, so huge a rout
Encumbered him with ruin; HELL at last
Yawning received them whole, and on them closed. . . .
Meanwhile the adversary of God and man,
Satan with thoughts inflamed of highest design,
Put on swift wings, and towards the GATES OF HELL
Explores his solitary flight; sometimes
He scouts the right hand coast, sometimes the left. . . .
HELL bounds high reaching to the horrid ROOF,
And thrice threefold the GATES; three-fold were brass,
Three iron, three of adamantine rock,
Impenetrable, impaled with circling fire,

Yet unconsumed. Before the GATES there sat
On either side a formidable shape.
The one seemed woman to the waist, and fair,
But ended foul in many a scaly fold.
Voluminous and vast, a serpent armed
With mortal sting; about her middle round
A cry of HELL HOUNDS never ceasing barked
With wide Cerberean mouths full loud and rung
A hideous peal; yet, when they list, would creep,
If ought disturbed their noise, into her womb.
And kennel there, yet there still barked and howled. . . .
The KEY to this infernal PIT by due,
And by command of heaven's all-powerful king,
I keep, by him forbidden to unlock
These a adamantine GATES; against all force
Death ready stands to interpose his dart,
Fearless to be o'er matched by living might. . . .
He rules a moment; CHAOS umpire sits,
And by decision more embroils the fray
By which he reigns; next him high arbiter
Chance governs all. Into this will ABYSS,
The WOMB OF NATURE and perhaps her grave,
Of neither sea, nor shore, nor air, nor fire. . . .
To whom Satan turning boldly, thus. 'Ye powers
And spirits of this NETHERMOST ABYSS,
CHAOS and ancient Night, I come no spy,
With purpose to explore or to disturb
The secrets of your realm, but by constraint

Wandering this DARKSOME DESERT, as my way
Lies through your SPACIOUS EMPIRE up to light. . . .
Deep to the roots of hell the gathered beach
They fastened, and the mole immense wrought on
Over the foaming deep high arched, a BRIDGE
Of length prodigious joining to the wall
Immovable of this now fenceless world
Forfeit to death; from hence a passage broad,
Smooth, easy, inoffensive down to HELL.
So, if great things to small may be compared,
Xerxes, the liberty of Greece to yoke,
From Susa his Memnonian palace high
Came to the sea, and over HELLSPONT
BRIDGING his way, Europe with Asia joined,
And scourged with many a stroke the indignant waves.
Now had they brought the work by wondrous art
Pontifical, a right of PENDENT ROCK
Over the VEXED ABYSS, following the track
Of Satan, to the selfsame place where he
First lighted from his wing, and landed safe
From out of CHAOS to the outside bare
Of this round with pins of adamant
And chains they made all fast, too fast they made
And durable; and now in little space
That confines met of empyrean heaven
And of this world, and on the left hand HELL
With long reach interposed; three several ways

In sight, to each of these three placed led."[156]

[156] *Paradise Lost and Other Perms* John Milton, Published for the Classics Club by Walter J. Black, Inc. NY, 1943, pp 132-3, 137-9, 141, 242, 321-322. Original poem was written around 1640.

CHAPTER 11
THE COMING RACE

THE "SAVIOR" CONCEPT:

In chapter 1 Blavatsky described the Inner Earth as: "the cradle of the first man and the dwelling of the last divine mortal, chosen as the FUTURE SEED OF HUMANITY."[157] De Lucia states in his book, *Hollow Earth in the Puranas*: "A the end of Kali Yuga, the Kalki AVATAR will be born in the best of Brahmin families of the city of Shamballa to ANNIHILATE MISCREANTS ON THE SURFACE of the globe. Afterwards, the general Puranic version goes that Vedic culture will be resettled on the surface of the Earth from the interior of the planet."[158] Does this mean that Krishna, of the BLUE RACE, will return? Krishna, born about 7,000 years ago, was an incarnation of Vishnu. The death of Krishna is said to have inaugurated the Kali Yuga.

Before examining the possibility of a "savior" for mankind, coming from the Inner Earth, and a super race that may "annihilate miscreants", let us focus upon the BLUE RACE. Is it possible that such a race exists? The following story was told by Jacques Bergier:

> "One afternoon in August, 1887, near the village of Banjos in Spain, some peasants working in a field saw two children come out of a cave, a boy and a girl whose clothes were made of a material they did not recognize and whose skin was as GREEN as the leaves on the trees. . . . Specialists from Barcelona tried in vain to identify the language they spoke and to

[157] *The Secret Doctrine,* Blavatsky, Vol. II, pp 6, 220.

[158] *Hollow Earth in the Puranas*, p 9.

analyze the material of their clothing. The children were remanded to a local justice of the peace, Ricardo da Calno. He tried to rub off their GREEN color, but it was not makeup it was the true pigmentation of their skin. The children's faces were noted to have certain negroid traits but their eyes were of a rather asiatic type, almond-shaped. For five days, the children were given a wide variety of foods, which they refused to eat. Finally, they were brought some freshly cooked beans, which they accepted. The boy, who was too far weakened, died. The girl survived. The GREEN color of her skin gradually faded, leaving her a normal complexion for a person of the white race. She learned a little Spanish, but when she was questioned, her statements only complicated the mystery. She described the country from which she came; a country without sun, where a permanent twilight reigned. This country was separated by a river from a bright country where the sun shone. There had been a whirlwind suddenly, accompanied by a terrible noise, and it had lifted the two children and deposited them in the cave. The young girl survived for five years before dying."[159]

A slightly different version of the same story, and a different story, about green children appearing in England in the twelfth century, was relayed by William Michael Mott.

---

[159] *Extra-Terrestrial Visitations from Prehistoric Times to the Present*, Jacques Bergier, Henry Regnery Company, IL, 1970, pp 127-130.

He says these reports were recorded in the *Chronicon Anglicarum* and *Historia Rerum Anglicarum* by chroniclers and monks Ralph of Coggeshall, William of Newbury, and Walsingham and Giraldus Cambrensis.[160] Timothy G. Beckley, in his book *Subterranean Worlds Inside Earth;* Thomas Keightley, in his *The Fairy Mythology,* published in 1850; John Macklin's *Strange Destinies*; W. R. Drake's article on "The GREEN PEOPLE"; and Warren Smith, in *Hidden Secrets of the Hollow Earth,* also wrote about this and similar incidents. Some authors claim the children were from "St. Martin's Land", an Inner Earth area named after the Saint who had visited them.

According to David Jungclaus, Ph.D., in his book *The City Beneath the Bermuda Triangle*, the ancient green race came from the Star Letsing.[161] The GREEN skin is particularly interesting since Krishna (Kris meaning sacred), was born "BLUE". Many people, who have seen beings from the Inner Earth, have reported that they had a BLUISH or GREENISH cast to their skin.

Doreal also wrote about the BLUE RACE, saying:

> "For over two hundred thousand years before Atlantis sank the BLUE RACE had lived in that interior part of earth. . . . They supplanted a yet more ancient race who were the guardians of the center of the Earth."[162]

---

[160] *Caverns, Cauldrons, and Concealed Creatures*, William Michael Mott, TGS-Hidden Mysteries, TX, 2002, pp 127-130.

[161] *The City Beneath the Bermuda Triangle* David Jungclaus, Ph.D, Lost Word Publishing, CA, 1989, p 14.

[162] *The Inner Earth*, Doreal, Brotherhood of the White Temple, CO, p 10.

"The spiritual Blue Race, which chose to live and work in the inner earth, was originally one of the races on Atlantis. Some of those races have changed, some have vanished utterly, and some still remain in small numbers. This Blue Race once inhabited part of the European continent, which is now the island of Great Britain. The ancient Britons and even down to historic times, painted themselves with blue clay or blue woad. They had some faint knowledge of a very ancient wise race which inhabited Britain before historic times. The Blue Race, which originally lived in Atlantis, and later in Britain, of its own choice, took up the task of guarding the inner earth spaces. That Blue Race evolved to the point where they did not need eyes.

They developed instead a sense of perception which was far superior to that of sight. . . . The first modern work that mentioned this Blue Race was written by an occultist, John Uri Lloyd, in his book *Etidorhpa*. He only spoke of the Blue Race symbolically, and gave certain information that until that time had not been released to anyone except initiates. . . . At this particular time [New Age] when the forces of destruction are rampant on this earth, contact between the Anti-Christ and those evil ones in prison has been established.

Therefore, the Blue Race, in the inner earth, that race which chose of their own will, to become the guardians of the inner earth spaces and of those evil ones

> bound in prison, are tried almost beyond their power of endurance, so that many of the great souls on this earth, those ('Warriors of the Dawn') who have passed into higher initiations, are working on holding and keeping closed that barrier."[163]

Gary Smith, in his article, "Shiva and Krishna, The Ascension of the BLUE RACE," states:

> "Many thousands of years ago in the Raja Pura Valley in India lived the BLUE RACE of humanity. . . . Only legend now remain. However, both Shiva and Krishna were from the BLUE RACE. Long ago, Shiva brought forth the knowledge of Creation. . . . Krishna was Shiva's student, and thus were born, the Vedas. . . . The BLUE RACE decided to ascend, however, a few were left behind to create their Shiva Mystery School."[164]

There was a race of Blue Moovians (Lemurians of the third root race) who were very tall, about seven feet, very thin, and had extremely large heard. They possessed all manners of powers of the mind, including teleportation, telekinesis, ESP. One day, in response to a stimulus known only to them, they conveyed to regular humans that they had, through their powers of astral projection, located a more suitable planet, and disappeared simultaneously.[165] There is

---

[163] *The Mysticism and Mysteries of the New Age*, Doreal, Brotherhood of the White Temple, Inc., CO, 1958, pp 9-13, 15.

[164] spiritualminds.com

[165] The Straight Dope: Is there really a race of blue people?, p 2 of 4; webmaster@straightdope.com; or cecil@chicagoreader.com

also a legend that the Ainu or Aini people, an ancient race living in Japan, were a pale shade of blue.

Some schools of thought consider the blue color to be an accumulation of negativity. If this be the case, Krishna is considered to have turned blue because he took on and transmuted the negativity of the world, just as Jesus transmuted world karma by the shedding of his blood. Other schools of thought believe the blue or green coloration is a result of particle radiation from living underground, away from the ultra-violet rays of the sun. Perhaps this is analogous to the change in coloration of transparent fish who live very deep within the ocean--too deep for the rays of the sun to penetrate.

Will a remnant of these BLUE or GREEN people return to the surface, along with the fair-skinned Aryans and Nordics who have promised to assist mankind by "annihilating the miscreants"? Dr. Ferdinand Ossendowski, in his book *Beasts, Men and Gods*, sites a remarkable prophecy given by the King of the World, Rigden-Jyepo, who lives in the Inner Earth Kingdom of Agharti. The prophecy was conveyed to him by Lama Hutuktu of Narabanchi in 1921. The Lama said the prophecy had been given around 1890:

> ". . . Then I shall SEND A PEOPLE, NOW UNKNOWN, which shall TEAR OUT THE WEEDS OF MADNESS AND VICE with a strong hand and will lead those who still remain faithful to the spirit of man in the fight against evil. They will FOUND A NEW LIFE ON EARTH purified by the death of nations. In the FIFTIETH YEAR only three great kingdoms will appear, which will exist happily SEVENTY-ONE YEARS.

> Afterwards there will be EIGHTEEN YEARS of war and destruction. THEN THE PEOPLES OF AGHARTI WILL COME UP from their subterranean caverns to the surface of the earth."[166]

The time frame on this prophecy would be approximately: "FIFTIETH YEAR" - from 1890 to 1940; "SEVENTY-ONE YEARS" from 1940 to 2011; "EIGHTEEN YEARS" from 2011 to 2029. However, there have been several errors in our calendars through the years, as well as changes in the timing of the prophecies themselves. The prophet Nostradamas had predicted the greatest holocaust to occur in 1999, then predicted a twenty-seven year war, which would bring the date of the New Age to around 2026. According to the ancient Mayan Calendar, the New Age is to begin in 2012. When dealing with prophecies, it is helpful to remember that they are meant only as warnings; if there is enough "light" and the state of consciousness changes, the prophecy will be altered accordingly.

During this time of war and great strife, Thoth, in *The Emerald Tablets*, says:

> "Weapons of force shall wipe out the earth-men until half of the races of men shall be gone. Then shall come forth the SONS OF THE MORNING. . . . Then shall the ancient home of my people RISE FROM ITS PLACE 'neath the dark ocean waves."[167]

---

[166] *Beasts, Men and Gods*, Dr. Ferdinand Ossendowski. This quote was taken from "A Prophecy from the Inner Earth", http://www.geocities.com/CapitolHill/6824/sacrgeo.htm

[167] *The Emerald Tablets of Thoth*, Tablet XII, p 48

This is when the Inner Earthers, along with Jesus and the benevolent sky people, will come to save mankind and our ecosystem, and usher in a New Age. According to the book, *The Coming Race,* by Lord Edward Bulwer Lytton, a SUPER-ADVANCED RACE from the Inner Earth plans to RETURN TO THE SURFACE to help humanity to adjust to the new-age frequencies. Lytton's book offers a BLUEPRINT FOR THE FUTURE with the motto: "No harmony without order; no order without authority, no authority without unity."

This super race, living in the earth's interior, have harnessed a unique energy which they call Vril, a word derived from virile. It is a power that has been known and studied by the Chinese for centuries. It has been called the electricity of nature, orgone, odic force, or atmospheric magnetism. It is also related to the energies of chi, feng shui, kundalini, shakti, melatonin, and chrism. According to Lytton, Vril has the ability to wash away memory, control the weather, heal all ailments, allow thought transference and radiate light that suppresses all other forms of light. He related that every member of the advanced "Vril-yans society carried a small tube of this milky white liquid". He describes:

> "There is no word in any language I know which is an exact synonym for VRIL. I should call it electricity, except that it comprehends in its manifold branches other forces of nature, to which, in our scientific nomenclature, different names are assigned, such as magnetism, galvanism, etc. These people consider that in VRIL they have arrived at the unity in natural energetic agencies, which has been conjectured by many philosophers above ground, and which Faraday thus intimates under the

more cautious term of correlation. . . . These subterranean philosophers assert that, by one operation of VRIL, which Farady would perhaps call 'atmospheric magnetism', they can influence the variations of temperature--in plain words, the weather; that by other operations, akin to those ascribed to mesmerism, electro-biology, odic force, etc, but applied scientifically through VRIL conductors, they can exercise influence over minds, and bodies, animals and vegetable.

This fluid is capable of being raised and disciplined into the mightiest agency over all forms of matter, animate or inanimate. It can destroy like the flash of lightning; yet, differently applied, it can replenish or invigorate life, heal, and preserve, and on it they chiefly rely for the cure of disease, or rather for enabling the physical organization to re-establish the due equilibrium of its natural powers, and thereby to cure itself. By this agency they rend way through the most solid substances, and open valleys for culture through the rocks of their subterranean wilderness. From it they extract the light which supplies their lamps, finding it steadier, softer, and healthier than the other inflammable materials they had formerly used."[168]

Blavatsky, explaining VRIL as being similar to "FOHAT", says:

---

[168] *The Coming Race* Lytton, pp 28-29, 33-34.

> "A Force, whose secret powers were thoroughly familiar to the ancient theurgists, is denied by modern skeptics. The antediluvian children--who perhaps played with it, using it as the boys in Bulwer Lytton's *Coming Race*, use the tremendous 'VRIL'--called it the 'Water of Phtha'; their descendants named it the Anima Mundi, the soul of the universe; and still later the medieval hermetics termed it 'sidereal light,' or the 'Milk of the Celestial Virgin,' and 'Magnes,' and many other names. But our modern learned men will neither accept nor recognize it under such appellations; for it pertains to magic, and magic is, conception, a disgraceful superstition."[169]

> "It is the terrible sidereal Force, known to, and named by the Atlanteans Mash-Mak, and by the Aryan Rishis in their Ashtar Vidya by a name that we do no like to give. It is the VRIL of Bulwer Lytton's *Coming Race*, and of the coming races of our mankind. The name VRIL may be a fiction; the Force itself is a fact doubted as little in India as the existence itself of the Rishis, since it is mentioned in all the secret works."[170]

In modern terminology, we might refer to VRIL as the harnessing of nuclear power or as a laser beam. According to Lytton, the "Vril-ya" race did not want their residents to know of progress being made on the surface, 'lest

---

[169] *Isis Unveiled, Vol. I*, Blavatsky, p 64.

[170] *The Secret Doctrine, Vol. I,* Blavatsky, p 563.

some of their inhabitants might venture there. They were afraid the surface visitor would leak the information to the "Vril-yans," or if allowed to return to the upper world, he would tell of their existence "down under". Therefore, they decided to put him to sleep permanently. Fortunately, one of the girls who had fallen in love with him and orchestrated his escape. Therefore, he lived to tell the tale of the "Vril-yans" in *The Coming Race.*

Perhaps when they come to the surface, we will finally learn to harness this great force for good instead of using it for the purposes of war and destruction. It sounds like those who disagree with the super-advanced race--those they term "MISCREANTS"--will be "ANNIHILATED". As the Tibetan prophecy stated: "I shall send a people which shall TEAR OUT THE WEEDS OF MADNESS AND VICE." There is certainly little doubt that many "weeds of madness and vice" exist today.

This "annihilation" or "tearing out" is explained in the Bible (Rev. 21:8).

The Bible also predicts the coming of people from the Inner Earth:

> "And I shall set up an ensign for the nations, and shall assemble the outcasts of Israel, and GATHER TOGETHER THE DISPERSED of Judah from the FOUR CORNERS OF THE EARTH [the Inner Earth is the only location that has 'Four Corners']." Isa. 11:12
>
> "I will say to the north, give up, and to the south, keep not back, BRING MY SONS FROM FAR, and MY DAUGHTERS FROM THE ENDS [interior] OF THE EARTH." Isa. 43:6

> "Art thou not it which hath dried the sea, the waters of the great deep; that hath made the depths of the sea a way for the RANSOMED TO PASS OVER." Isa. 51:10

> "Thus saith the Lord, a PEOPLE COMETH FROM THE NORTH COUNTRY, and A GREAT NATION SHALL BE RAISED from the SIDES [inside] OF THE EARTH". Jer. 3:18; 6:23.

We are blessed that Divine Intervention is standing by to help us through the planet's final tests. Not only will (BLUE) Krishna and other ENLIGHTENED BEINGS from the Inner Earth come to help mankind through its times of "tribulation", but JESUS himself will come in great CLOUDS (UFOs) of glory. The Bible says:

> "That in the AGES TO COME [the Aquarian Age] He [God] might shew the exceeding riches of his grace in his kindness toward us through Christ Jesus." Eph. 2:7

> "And then shall appear the sign of the SON OF MAN in heaven: and then shall all the tribes of the earth mourn, and they shall see the Son of man coming in the CLOUDS of heaven with power and great glory. And he shall send his angels with a great sound of a trumpet, and they shall GATHER TOGETHER his elect from the FOUR WINDS, from ONE END OF HEAVEN TO THE OTHER." Matt. 24:30-31

> "For the SON OF MAN shall come in the glory of his Father with his angels; and then he shall reward every man according to his works." Matt. 16:27; 2 Thes. 1:7; Rev. 22:12

> "Behold, he cometh with CLOUDS; and every eye shall see him, and they also which pierced him; and all kindreds of the earth shall wail because of him." Rev. 1:7
>
> "And Enoch also, the seventh from Adam, prophesied of these, saying, Behold, the Lord COMETH WITH TEN THOUSANDS OF HIS SAINTS to execute judgment upon all." Jude 14-15
>
> "When the SON OF MAN shall COME IN HIS GLORY, and all the holy angels with him, then shall he sit upon the throne of his glory: and before him shall be gathered all nations; and he shall separate them, one from another, as a shepherd divideth his sheep from the goats: and he shall set the sheep on his right hand, but the goats on the left.
>
> Then shall the king say unto them on his right hand, Come ye blessed of my Father, inherit the KINGDOM PREPARED for you from the foundation of the world." Matt. 25:31-34

Jesus said: "And I, if I be LIFTED UP, will draw all men unto me." John 12:32 Will we be BEAMED UP into a UFO, the "CLOUDS of Jesus", during the "rapture", as a holding tank, then taken DOWN into the Inner Earth? According to the Bible, we will live in "Paradise" [Inner Earth], the "place" that Jesus went to "prepare for us". John 14:2; Rev. 21:10 Is the Inner Earth the "New Jerusalem" or is "New Jerusalem" a huge mothership used for transporting? Or both? Whatever arrangements are being made, according to the Bible, we are to dwell there in peace and harmony, with Christ, for a THOUSAND YEARS, during which time "Satan" will be bound. Rev. 20:2-6 The *Book of Enoch* agrees (see Enoch 39:4-10; 41:1-2; 92:14). Perhaps it will take a thousand years to cleanse the surface of the earth from the debris of war, especially if there is nuclear fallout, massive volcanic action, or an ice age, or some of each.

Will we retain our same form, or will our physical bodies be CHANGED when we are taken inside the Inner Earth? Evidently there will be some CHANGES. According to Mack, in his book *Abductions*, even the "little greys" will be of help. One of his clients revealed, while under hypnosis:

> "Lines were being drawn, spokes or tunnels through which people would pass into the future. There will be a shift in time, and it will be like a veil that people are going to walk through. . . . The species of man will continue, but in a different form. The earth is going TO OPEN UP TO A PLACE [Inner Earth] OF INTERACTION with other beings, including hybrid [alien] children now being created. Once the humans who are part of this plan move through time and through the shift of consciousness, there will be no anger or hatred or resentment among the abductees who are shown their children . . . That which you feel is cruel is just the human mind trying to understand."[171]

Concerning this change, the Bible says:

> "Jesus Christ who shall CHANGE OUR VILE BODY, that it may be FASHIONED

---

[171] *Abduction, Human Encounters with Aliens* John E. Mack, M.D., Charles Scribner's Sons, Macmillan Publishing Co., NY, p 328.

> LIKE UNTO HIS GLORIOUS BODY, according to the working whereby he is able even to subdue all things unto himself." Phlp. 3:21
>
> "Beloved, now WE ARE THE SONS OF GOD, and it doth not yet appear what we shall be; but we know that, when he shall appear, WE WILL BE LIKE HIM; for we shall see him as he is." 1 John 3:2;
>
> "Behold, I shew you a mystery; We shall not all sleep, but we shall all be CHANGED. In a moment, in the twinkling of an eye, at the first trump: for the trumpet shall sound, and the dead shall be raised incorruptible and we shall be CHANGED" 1 Cor. 15:51-52

After we are "CHANGED", will our bodies radiate light like Jesus? The Bible promises:

> "Then shall the righteous [the saints] SHINE FORTH AS THE SUN." Dan. 12:2-3; Matt. 13:43
>
> "He that overcometh, the same shall be clothed in WHITE RAIMENT." Rev. 3:5
>
> "After this I beheld, and, lo, a great multitude, which no man could number, of all nations and kindreds, and people, and tongues, stood before the throne, and before the Lamb, clothed with WHITE ROBES, and palms in their hands." 7:9

Sadek Adam makes an interesting analysis in his book, *Hollow Earth Authentic*:

> "The body we now possess is an aberration of the normal, 'spirit' body

> resident in Paradise. This normal body is the one we occupy after death or develop in transmutation. It is physical, incorruptible, does not age/die/get diseased and has some parts missing and others added. It is interesting that UFO contactees speak of the blond UFO owner/occupiers as having extra vertebrae. Percival in his book also mentions that the immortal body has a second spinal cord. the details in this book are similar to those in *Here and Hereafter* - no generative and digestive systems; instead a mass of nerves and brains. Blood is a vital force. If we were to see an inhabitant of the hollow earth, she/he would look extremely healthy, vibrant, angelic, exceptionally beautiful/handsome, with a blemish-free skin and giving off an inner light, and no more than about 28 years old."[172]

This eternal youthfulness sounds like that of Princess Sharula Aurora Dux from Telos, as mentioned in chapter 7.

Whether we are to be taken into the holy subterranean metropolis of the Inner Earth "Paradise", or beamed into a spacecraft called "New Jerusalem", we will not see the sun. It will be divinely lighted. Could this be the meaning of the biblical scripture:

> "The sun shall no more be thy light by day; neither for brightness shall the moon give light unto thee; but the LORD shall be unto thee an EVERLASTING LIGHT, and thy God thy glory." Isa. 60:19

---

[172] *Hollow Earth Authentic*, Adam, p 91; *Thinking and Destiny*, Percival; *Here and Hereafter*, Ruth Montgomery.

> "And there shall be no night there; and they need no candle, neither light of the sun: for the Lord God giveth them light."
> Rev. 22:5

Then after the one-thousand-year period is completed, we will evidently be brought back to the surface of the earth that by then will have been cleansed and purified. The Bible says it will be a "new earth with no death, sorrow, crying or pain"--that "the old heaven and earth has passed away". Rev. 21 But also, at that time, "Satan" will again be released. Rev. 20:3, 7-15 Why? By then, we will have learned from our millennial sojourn into the Inner Earth, or our residence in a spacecraft, to overcome evil and live in peace. At least this is the divine plan. As long as we are an individualized spark of God, no matter what form we take, we will always be required to choose good over evil. Even the angels in heaven had the freedom to rebel. But with the "miscreants annhiliated" and our thousand-year sojourn in the paradisiacal conditions, we will certainly have every opportunity to remain positive and choose to walk in the light. In the immortal words of Thoth:

> "Then shall the Age of Light be unfolded. . . . Then shall the Brothers of Light rule the people. Banished shall be the darkness of night."[173]

---

[173] *The Emerald Tablets of Thoth*, Tablet XII, p 48.

CHAPTER 12
PERSONAL EXPERIENCES

THE HANDS-ON CONCEPT:

KATRINA HILL'S EXPERIENCE WITH THE "BLUE PEOPLE":

"My first experience was when I was camping at Crestone, Colorado, I meditated near the base of the mountain at about 10,000 feet elevation. As I was guided to fix my attention on a particular section of rocks, two blue beings slowly came through them, as if the stones were made of water or a gaseous substance. It seemed as if the beings just blended with the rocks, then emerged from them. There was evidently some kind of a molecular interchange. I was at total peace as they radiated incredible love, the pure energy of God. I felt with my heart in some type of soul realization. It was the purest energy I had ever experienced with no thoughts or wavelength. Although they didn't say a word, their presence brought a change in my consciousness. I felt intense gratitude.

"They were of a bluish-turquoise color. I couldn't see through them, but knew they were in some type of a light body. They had arms, hands, and legs, but I saw no feet walking. They just more or less floated around. I did not see their faces in a normal way, but saw from inside my mind instead of with my eyes. My entire perspective was changed. The energy was so high I wondered if I had switched dimensions. I knew that words were unnecessary because it's not about explaining, but just about being.

"While I was in Crestone, I saw them on three occasions. Each time was similar. There was never any question in my mind about their reality. I have not been the same since they increased my frequency. I was definitely graced by their appearance.

"My second experience was at Mount Ashland, Oregon. I had come there with a friend to camp during the full moon. We stayed up nearly all night, watching it move across the sky. In the early dawn, I looked across to an outcropping of rocks. I was fascinated with the rock formations, and began to focus intensely. Suddenly, I saw a perfect doorway forming a huge arch. A section of the mountain seemed to open up. Two beings emerged and turned sideways to look across at us. My friend was asleep. These beings were blue-turquoise, as were the ones in Crestone, but were of a much paler shade, a type of white-blue. It seemed as if they were robed in a white-crystalline light and that their turquoise bodies were shining through it. Again, there were no words, but an intense radiation of pure love, and again, I did not question their reality.

"During my third encounter, I did not see the beings. My first two experiences had been felt on a heart/soul level, but this was a sound attunement. I had been guided to walk a labyrinth in Ashland, Oregon on Good Friday, Saturday, and Easter Sunday. While walking it on Sunday, I was attuned vibrationally until I began hearing or sensing a sound vibration from the Inner Earth. It was a frequency rather than a verbiage. I realized that it was a TUNE that allows one a deeper atTUNEment. I also sensed magnificent colors and was bathed in them. Without words, my inner knowing revealed: "The Inner Earth beings are beginning to sing their song, and we are just now beginning to hear it. In this way, mankind will be uplifted." Katrina Hill

(Note from author: The Book of Revelations states: "And they sang a new song." Rev. 5:9; 14:3.)

Katrina Hill is a spiritual teacher, author and poet. With her extraordinary energy and talents, her friends all consider her to be an angel who has descended to help humanity. When I asker her what to say about her classes, teachings, writings, etc., she said: "Just say: To be the living presence of love in form, and to give that love is not about who am I or how am I. I don't need something to validate me, I just need to be that love in action. Tell others to be the love that they are."

CLARA LEDBETTER'S EXPERIENCE WITH THE TINY BEINGS IN THE INNER EARTH:

"On a summer day in 1985, I was very ill with an internal infection and was going in and out of consciousness with periods of sleep and awake.

"During one of these short sleep spells, I found myself squatting beside a small stream running through a path which led to the kitchen area. While playing in the water, my friend came to ask me if I would like a bowl of chicken soup; my body was still in the bed. While doing so, he walked right through me as I was beside the water in my spiritual body. After he asked me about the soup, he returned and walked through me again. I thought this odd as I had heard him ask me about the soup and also my answer from the bed.

"I was attracted to some mica shimmering in beautiful colors in the water. I tried to pick up the tiny flecks and place them together by color on the rocks. I had arranged several pieces according to color when two little gnomes walked by, close to me. They watched me for a few seconds then one of them asked if I knew what it was I was gathering. They were only eight or nine inches tall. I told them it was mica that always shined so beautifully in the clear bubbling water.

"They took turns explaining to me that they used the

mica for healing. This really perked my interest. I was impressed by the idea which I had never heard before.

"After a few minutes of visiting, they asked me if I'd like to see how this was done. I was really excited by this time. I gave them my enthusiastic 'Yes, I'd love to.'

"One of them asked, 'Will you go now?'

"My reply was. 'Sure, I would be thrilled to.'

"Before I knew what was taking place, I was small like them, around nine inches or so tall. Then we were on our way.

"We went underground, literally flying through tunnels, places like long caves, always turning and zipping around solid objects such as rocks, tree roots and such, so as not to disturb nature. As we traveled, I was carried between these two little fellows. I never touched the ground during the whole trip.

"In just a little while we arrived in a huge cave-like place where people were preparing for a secret place of safety in the future. Here, they already had two elevator shafts in place and one laying on its side. We merely jumped to the top side.

"To my surprise, the place became bright and filled with light, but I could not distinguish where it came from. Then I noticed a huge pile of mica. All of the rainbow colors had been laid right in the piles on the cylinder that we were standing on. They explained to me that the workers didn't know the mica was there as it was in another dimension of space. I didn't understand this until later on.

"From here, we traveled again through the ground and then surfaced on a sloping hillside covered with small brushy bushes. There were trees of which I'd never seen before, and a gently flowing stream circling around the area.

"A gentle breeze began to make the trees wave in and blow as if greeting us ever so gently. During this time, thousands of elves, fairies, gnomes, etc. had gathered to the area and joined us. They sat on the branches and held on to the smaller limbs, with an expectant, happy enthusiasm that was a real joy and pleasure to witness.

"Within a few minutes, the ones in charge, called forth for the mica to begin floating on the breeze. The sunlight reflected off the shinning mica much like tilting a mirror to send the light to bathe each of us in an electrifying, pulsing, wave of energy that penetrated our very souls as well as our bodies.

"Each color took its turn, lasting a few minutes, with some longer, until everyone was cleansed, healed and full of joy that was different from anything I'd ever witnessed before.

"The ones called forth to the vibration relating to the color like leaves answering to the call of a fall breeze. My two companions told me that all the mica except the few pieces I was attracted to in the little stream would be recalled back into the piles according to its vibratory rate, each one had its own vibration to answer and respond to.

"Try as hard as I can, not being a writer, I cannot express the whole body feeling of being completely showered, through and through with the vibes as waves of pure radiance of love, peace, joy and wellness that I felt. This was my most delightful trip through the years and I've remained in good health ever since."

Clara, now in her 80's, has been reading, studying and

walking the spiritual path for many years. She is interested in UFOs and is against the government "cover ups". Her encounter with the little gnomes is only one of her many experiences.

JORGIE MARIE'S DREAM EXPERIENCE IN THE INNER EARTH:

"It was in May of 1981 when this experience happened. I know the date because I had been recording and dream journaling. Not that I had to read it for clear recall--the memory has been planted in my mind's eye as if it happened yesterday. So for me, it is a pleasure to share this dream in detail. It took place in a time that I had been meditating and achieving high ascensions. Lucid dreaming had allowed me to become one with myself. Fear was no longer on the menu.

"The dream begins as my friends are driving me to a place in town called The Western Plaza. We were to meet there and share an afternoon with a lunch-time discussion. To my surprise, the plaza had set up outside tables with umbrellas, just perfect for the occasion. Most of the shops had set up flea-market-type booths and had drawn many people. As I sat very relaxed, observing the crowds, I heard a noise from above in the clear blue sky. I looked up and saw hundreds of lights falling toward earth. I felt an energy was coming our way. As a flash of remembrance came across my mind, I turned to ask someone the time. He said it was 10:00. At this point I knew I was dreaming. It was a future event and I knew the time and date of its occurrence. The funny thing was that I knew I would forget the date and only remember the time.

"Then the earth began to shake violently. Shops were tumbling and the sky was turning dark. Fire had broken out. People were running like many ants but mostly away from the lights that were coming closer. I had no fear, but only welcoming thoughts of "Wow, they are coming". I watched with anticipation and interest until one of my friends called my name and said, 'Jorgie, don't just stand there. Run!' I couldn't figure where they were running to. I answered, 'There is nothing to be afraid of'. She grabbed my arm and pulled me into the flow of the panicked people. I didn't resist her, but thought, 'This is where the danger is, in the fear stampede'. I turned to watch the lights, as she pulled me forward.

"They were more than lights now, they were spacecrafts landing and hovering over the plaza. I smiled, with a surge excitement, then was a period of blackness. It was comfortable, but no telling exactly how long the time frame was. In dream time, it seemed like thirty seconds.

"My next point of awareness was in a very large auditorium made of stone. The size could have been as large as a football field. It was dimly lit except for the entertainment on the stage. There, moving colors of light and colored water was flowing and falling while chosen beings were dancing to the rhythmetic sound that the water made. It seemed to be a method of keeping the focus of the crowd front and center. The ones dancing on stage I recognized from dreams of space beings as a child. I also felt strong love connections; the feeling was akin to 'this is my family'. But more than just observing what was around me, I had an inner feeling of belonging, the feeling of positive fullness, wholeness, and the exhilaration of wellness. It was close to impossible to have a negative thought in this place. A thought of my mother entered my mind. Where was she? A voice covered the thought saying, 'She is in a safe place'.

"My attention returned to the beings on the stage and the color light and water show. There, sitting in my seat, my over enthusiasm drew me into being part of it. This thought had brought the water under my feet, and it was splashing around me as I was keeping time with the rhythmic sounds. I laughed and noticed that my friend next to me was getting wet from my actions. She giggled and motioned for me to calm down a bit. I pulled in my own energy field so as not to disrupt her enjoyment.

"My senses were keen and I began to notice things around me. I was on an isle seat and could see the wall. It was made of enormous rocks stacked on top of each other. I was sitting across from a space which made an entrance into the stone wall, and from this space I heard cracking sounds. This didn't disturb me but it did arouse my curiosity. There was a man walking slowly down the isle towards me with a flashlight object that I would call a laser light, only much more advanced. As he came closer, I had flashes of a memory from a long time ago in earth time, and recognized him as someone I called my brother. He had instructed me via dreams and we had spent many hours together. I felt very much one with him.

"He noticed that I had been looking over the large crack entrance in the wall and came over to ask me if I had heard anything. I answered 'Yes,' and to that he went to inspect the noise with his light. A short time passed and he returned. When he came out I asked him what was causing the sounds. He said, matter of factly, 'It's the expansion and contractions of the outer earth until it settles down.' It wasn't until this time that I realized and blurted out, 'You mean to tell me we are in the middle of the earth?' He looked at me with surprise, probably wondering how I had forgotten. 'Well yes,' he answered. I sat pondering the idea. Before this, I had no clue that this place existed. He was watching for my reactions. I simply couldn't conjure up any fear or negativity, but I did have curiosity. I asked him, 'Does this mean that I won't see the sun or hear the ocean? How long will this take?'

"At this thought, immediately, without blacking out, I was on the ocean in a little motor boat. The sky was big and blue, but it had no sun. The same wonderful feeling was still with me. I gazed around and saw that I was just off shore from a beautiful crystal city. There was a pier that jetted out from it and a larger ship was docked at the end of this pier. The voice of my brother said, 'Do you see what I mean?' I certainly did--your thoughts shall make it happen. This was all and everything that I had ever imagined. 'Thank you', I said to him, 'I think I will go visit the city'.

"I slowly placed my hand on the motor to energize the boat, then steered it over to the pier and tied it off. I climbed onto the pier. My first impression was one of amazement. Even the pier was made of crystal. Such freedom was radiating from within me. There was nothing I needed or wanted. All things were for enjoyment of the beauty and simplicity of being. I slowly walked off the dock and onto the cobblestone crystal street. There were shops to enjoy all sorts of creative ideas. Nothing was for sale because there was no need to possess anything, only to enjoy it. The feel of creativity was in the air. I walked outside and headed down the sidewalk to enjoy more sights.

"Just at that moment an adult giraffe was walking towards me, but he was only the size of a two-foot-high stuffed animal. He had stopped and was rubbing his back on the brick wall of the store front. I was in awe of how cute he was. I thought, 'You are so cute'. He looked over to me and thought back, 'You are cute too'. All thoughts of the mind are wide open here.

"I walked further and was now greeted by a small white kitten with bounding energy. She said, by her mind, 'Follow me'. Not only could I hear the thought, but the intentions behind the words, and this kitten wanted me to play. I picked up my steps to keep up with the playful romping of the kitten. She went a few buildings down and turned into an alleyway, and then again into another one. We were between large skyscrapers. Here in the alley was the kitten's mother. To my surprise this mother cat was ten feet high and ready to attack. This was a fine test of will and insight, testing my fear levels. I remembered clearly my dream teachings--when meeting an adversary offer a gift. At this time, I was lucid, so materialized in my closed hand a piece of bread soaked in milk. I extended my arm to this huge cat with the feeling of trust behind my intentions, and opened my hand. There, it was ready for the offering. She took it with gentle care and turned to take it back to her place of comfort. The little kitten seemed to be smiling. She jumped with all fours into the air and we began to roll and tumble. The more I played the more I became a kitten until I had shapeshifted into one and sunk into the abandoned feeling of freedom of playtime.

I must have been rolling in bed because my husband gave me a kick and I awoke in my body, sorry to have been brought back so abruptly.

This was only one of the adventures of Inner Earth, but it was the first that I remembered. And for me, it's only a thought away. We are in the universe as the universe is also in us." Jorgie Marie

Who is Jorgie Marie? If one were to describe her, the topic of dreams would be the first attribute discussed. She has been an active, conscious, lucid dreamer since childhood. Now, at the age of 57, she has held dream workshops and discussion groups, and has been journaling her own dreams on a nightly basis for more than twenty-five years. If you ask her why dreams are so important, she would gladly explain that life experiences are your best teachers; that conscious dreaming gives you the eye of the soul to guide your choices.

She says: "There isn't anything unsolved with the gift of conscious dreaming and a heart of compassion."

Some of her choices in life have been a balanced blend of being a loving mother and grandmother, a meditator, a massage therapist, a minister, a teacher, an automatic channeler, a writer and poet, a shaman, a lover of nature, and an avid gardener. She is a firm believer in all the healing arts, and practices keeping a balanced frame of mind with a good sense of humor. She believes that to laugh at yourself and to feel the abundance of life on all levels is truly a gift in itself. And so it is.

I will leave you with this thought - perhaps the interior sun of the Inner Earth is analogous to our kundalini chakra that slowly rises upward to the heart chakra (Shamballa) to create a paradisiacal setting, then gains in momentum to continue rising up to purify the upper chakras (surface of the earth), and finally blazes out through the crown chakra to enter "heaven". Perhaps "walking in the light" leads to ENLIGHTENMENT! And as each individual consciousness achieves the critical mass of ascension, it will propel mankind into a new age or a new paradigm shift that will eventually manifest in a mass ascension. "As above so below; as within so without."

BIBLIOGRAPHY

Adam, Sadek, *Hollow Earth Authentic*, Health Research, WA, 1999.

Armstrong, Virgil "Posty", *ETs and UFO - They Need us, we don't Need Them.*

Barker, Gary, *They Knew Too Much About Flying Saucers* Saucerian Press, WV, 1975

Beckley, Timothy Green, *Subterranean Worlds Inside Earth,* Inner Light Publications, 1992. (*)

Bergier, Jacques, *Extra-Terrestrial Visitations from Prehistoric Times to the Present*, Henry Regnery Company, IL, 1970

Benton, Floria, *Hollow Earth Mysteries and the Polar Shift.* Saiceroam {ress Omc/. WV. 1981.

Bernard, Raymond, *Agharta, The Subterranean World,* Fieldcrestr Publishing/Carol Publishing Group, 1964.

Bernard, Dr. Raymond, *The Hollow Earth, The Greatest Geographical Discovery in History,* University Books Inc., Lyle Stuart Inc., 1969; 1979 edition published by Bell Publishing Co., NY.(*)

Bletzer, June G. Ph.D., *The Donning International Encyclopedic Psychic Dictionary*, June G. Bletzer, Ph.D, Whitford Press, PA, 1986.

Bord, Janet, *Fairies: Real Encounters with Little People,* Carroll and Graff Publishers, NY, 1997.

Boschke, F. L., *The Unexplained, Mysteries surrounding Mt.*

Rainier.

Brian II, William L., *Moongate: Suppressed Findings of the U. S. Space Program,* Future Science Research Publishing Co., Or, 1982.

Bryson, Bernarda, *Gilgamesh, Man's First Story*, Holt, Rinehardt and Wiston, NY

Byrd, Admiral Richard, *A Flight to the Land Beyond the North Pole, The Missing Diary of Admiral Richard E. Byrd,* Inner Light Publications, Box 753, New Brunswick, NJ 08903.

Carter, Joseph H., *The Ultimate Reality*

Cater, Joseph H., *The Awesome Life Force* Cadake Industries, Fl.

Cavandish, Marshall, *Man, Myth and Magic, the Illustrated Encyclopedia of Mythology, Religion, and the Unknown,* NY, 1985.

Chaney, Dr. Earlyne, *Revelations of Things to Come,* Astara, CA, 1982.

Churchward, James, *The Lost Continent of Mu.*(*)

Clark, Jerome, *Encyclopedia of Strange and Unexplained Physical Phenomena,* Gale Research Inc., Wssh. D.C., 1993, pp 151-154

Collins, Andrew, *From the Ashes of Angels, the Forbidden Legacy of a Fallen Race,* Signet Books, London, 1997,

Commander X, *Underground Alien Bases,* Abelard Press, NY. (*)

Correlli, Marie, *Romance of Two Worlds,* Rand McNlly and Co., IL, 1886.

Crabb, Riley, *The Reality of the Underground,* Borderland Sciences Research Associates, CA, 1960

Dante, *Dante's Inferno.*

Davault, Paul, *On the Face of the Deep.*

De Lucia, Dean Dominic, *Hollow Earth in the Puranas*, TGS-Hidden Mysteries, TX, 2003. (*)

Dickhoff, Robert Ernst, Ph.D., *Agharta,* Bruce Humpheries, Inc. Publishers, USA, 1951.

Doreal, Maurice, *The Emerald Tablets of Thoth the Atlantean,* Brotherhood of the White Temple, Inc. Co, 1939.

Dyson, John, *The Hot Arctic,* 1979.

Emerson, Willis George, *The Smokey God,* Amherst Press, WI.(*)

Fitch, Theodore, *Our Paradise Inside the Earth,* IA, 1960.

Forster, Gerry, "From Mu to Thule and the Inner Earth," (article), 2001.

Giannini, Amadeo, *World Beyond the Poles*, Amherst Press.

Gilliland, James, *Becoming Gods; A Reunion with Source,* Self Mastery Earth Institute, OR.

Godwin, Joscelyn, *Arktos, the Polar Myth,* Adventures Unlimited Press, 1996.

Godwin, Joscelyn, *The Theosophical Enlightenment and Arktos: The Polar Myth.*

Greely, Adolphus W., *Three Years of Arctic Service* Charles Scriber's Sons, 1886.

Greer, Steven M., *Disclosure*, Carden Jennings Publishing Col, VA, 2001.

Hall, Manly P., *The Secret Teachings of All Ages*, Philosop[hical Research Society, 3910 Los Feliz Blvd, Los Angeles, CA 90027, 1977.(*)

Hall, Manly P., *Man Grand Symbol of the Mysteries,* Philosophical Research Society, CA, 1972.

Halliday, William, *Adventure is Underground.*

Hamilton, William, *Alien Magic.*

Hamilton, William, *Cosmic Top Secret.*

Hapgood, Charles Hl, *The Path of the Pole,* Chilton Book Company, 1970.

Herbert, Wally, *The Noose of Laurels.*

Horn, Dr. Arthur, *Humanity's Extraterrestrial Origin, ET Influence of Humankind's Biological and Cultural Evolution,* A & L Horn, CA, 1994. (*)

Howard, Dana, *Vesta, the Earthborn Venusian,* Essence Press, TX, 1950.

Jackson, Danny P. *The Epic of Gilgamesh*, Bolchazy-Carducci Publication, IL, 1997.

Kafton-Minkle, Walter, *Subterranean Worlds: 100,000 Years of Dragons, Dwarfs, the Dead, Lost Races, and UFOs from Inside the Earth,* Loompanics Unlimited, WA, 1989.

Lamprecht, Jan, *Hollow Planets, A Feasibility Study of Possible Hollow Worlds.* Armstrong Printing, TX, 1998.(*)

Lee, Bourke, *Death Valley Men,* MacMilan Co., NY, 1932.

Lee, Desmond, translated by, *Timaeus and Critias*, Penguin, NY, 1971.

Leedom, Tim Cl, *The Book Your Church Doesn't Want You to Read,* Kendall/Hunt Publishing, IO, 1993.(*)

LePoer Trench, Brinsley, *Secret of the Ages, UFO's From Within the Earth*, Pinnacle Books, NY, 1974.

Lewis, David H., *The Incredible Cities of Inner Earth*, Science Research Publishing House, 1979.(*)

Lloyd, John Uri, *Etidorhpa*, Pocket Books, NY, 1976, originally published in 1908..(*)

Luk, A. D. K., ***Law of Life, Book 2,*** A. D. K. Luk Publications, Ok, 1960.

Luk, A. D. K., *Law of Life and Teachings by Divine Beings, Book 1 and 2,* A. D. K. Luk Publications, Ok, 1978.

Lytton, Lord Edward Bulwer, *Vril - The Power of the Coming Race* Rudolph Steiner Publications, NY, 1972, original published in 1871.(*)

MacLellan, Alec, *The Lost World of Agharti, The Mystery of Vril Power,* Souvenier Press, London, 1982.(*)

MacLellan, Alec, *The Hollow Earty Enigma*, Souvenier Press,, Ltd., 43 Great Russel Street, London, 1999.(*)

McComas, Taylor, and Lama Choedak Yuthok, *Clear Mirror on Royal Genealogy*, Snow Lion Publications, NY, 1996

Menger, Howard and Connie, *The High Bridge Incident: The Story Behind the Story - Released After 35 Years of Silence*, Howard Menger, Fl, 1991.

Merritt, Abraham, *The Face in the Abyss* MacMillan, 1992.

Milton, John, *Paradise Lost and Other Poems*, Published for the Classics Club by Walter J. Black, Inc. NY, 1943.

Mott, Wm. Michael, *Caverns, Cauldrons, and Concealed Creatures*, TGS-Hidden Mysteries, TX, 2003. (*)

Nansen, Dr. Fridtjof, *Farthest North*, Harper and Brothers, 1896.

Neihardt, John G., *Black Elk Speaks*, University of Nebraska Press, Ne, 1932.

Norman, Eric, *The Under-people,* Award Books, NY, 1969.

Norman, Erik, *This Hollow Earth*, Lancer Books, NY, 1972. (See Warren Smith)

O'brien, Christine and Joy O'brien, *The Genesis of the Few, the Story of Those who Founded the Garden of Eden,* Turnstone Press, Wellingborough, England, 1985.

Oliver, Fredrich, ("Phylos"), *Dweller on Two Planets,* Borden Publishing Company, USA, 1952.

Ossendowsky, Ferdinand, *Beasts, Men and Gods,* E. P. Dutton and Co. NY, 1922.

Pauwels, Louis and Bergier, Jacques, *The Morning of the Magicians.*

Payne, Dr. Buryl, *Body Magnetic.*

Peary, R. E., *Northward*, F. A. Stokes Co., Ny, 1908

Picknett, Lynn and Prince, Clive, *Stargate Conspiracy,* Berkley Books, NY, 2001.

Rasmussen, Knud, *The People of the Polar North.*

Reed, William, *Phantom of the Poles,* Fieldcrest Publishing Company, NY, 1906.

Robbins, Dianne, *We are not Alone: Messages from the Hollow Earth and the Subterranean City of Telos,* Mt. Shasta Light Publishing, CA, 2000 (*)

Robbins, Dianne, *Telos, the call goes out from the Hollow Earth,* Mt. Shasta Light Publishing, CA. (*)

Robbins, Dianne, *Lailel,* Mt. Shasta Light Publishing, CA, 2000.

Roberts, Jane, *Seth Speaks*, New World Library, Amber-Allen Publishing, CA, 1972

Roerich, Helena, *Agni Yoga.*

Roerich, Nicholas, *Shamballa,* Frederick A. Stokes Co., NY, 1930.

Sanderson, Ivan T., *Invisible Residents,* The World Publishing Co., NY, 1970.

Shaver, Richard and Childers, David Hatcher, *Lost Continents and the Hollow Earth,* Adventures Unlimited Press, 1999.

Sitchin, Zecharia, *Divine Encounters,* Avon, NY, 1995.(*)

Sitchin, Zecharia, *The Stairway to Heaven*, Avon, NY, 1980.(*)

Smith, Warren, *This Hollow Earth*, Lancer Books, NY, 1972. (Published under his real naame Eric Norman)

Smith, Warren, *The Hidden Secrets of the Hollow Earth, Zebra Books,* Kensington Publishing, Inc., 1976.

Steiger, Brad, *The Hollow Earth: Myth or Reality.*

Steiner, Rudolph, *Cosmic Memory: Atlantis and Lemuria,* 1923.

Thompson, Dr. Richard L, *Alien Identities*, Govardhan Hill Publishing, 1993. (*)

Thompson, Dr., *Firbidden Archaeology*, Govardhan Hill Publishing, 2000. (*)

Tomas, Andrew, *On the Shores of Endless Worlds,* Souvenir Press, Ltd. England, 1974.

Time-Life Books, *Mystic Places*, VA.

Trench, Brinsley le Poer (Lord Clancarty), *Secret of the Ages: UFOs from Inside the Earth,* Souvenir Press, London, 1974.

Velikovsky, *Mankind in Denial.*

Velikovsky, Immanuel, *Earth in Upheaveal*, Dell Publishing Col, NY, 1955.

Velikovsky, Immanuel, ---------------, *Sphere Books Inc., London, 1950.*

Verschur, G. L., Impact, Oxford University, 1996.

Villars, Abbe N. de Montfaucon De, *Comte De Gabalis* The Brothers, London, 1670

Wagner, Richard, *The Gods Entry into Valhalla.*

Walton, Bruce, *A Guide to the Inner Earth,* Health Research, CA, 1983.

Walton, Bruce, *Mount Shasta - Home of the Ancients,* Health Research, CA, 1985.

Warren, William F., *Paradise Found, The Cradle of the Human Race at the North Pole,* Fieldcrest Publishing/Carol Publishing Group.

Wilkins, Harold T., *American Indian Myths and Mysteries*, Chilton Boods, PA, 1946.

X, Michael, *Rainbow City and the Inner Earth People,* Futura, CA, 1960.

Esoteric writings of Blavatsky, Prucker, Judge and Sinnett: Blavatsky, H. P. Collected Writings, Theosophy Publishing House, 1950-1991.

Blavatsky, The Inner Group Teachings of H.P. Blavatsky, H.

J. Spierenburm (comp.) Point Lama Publishing, 2nd ed., 1995.

Blavatsky, Isis Unveiled, Theosophy University Press, 1972 (1877).

Blavatsky, Theosophical Glossary, Theosophy Co., 1973 (1892).

Prucker, G. de, Dialogues of G. de Prucker, Theosophy University Press, 1948.

Prucker, G. de, The Esoteric Tradition, Theosophy University Press, 2nd ed., 1940.

Prucker, G. de, Wind of the Spirit, Theosophy University Press, 2nd ed., 1984.

Prucker, G. de, Fundamentals of the Esoteric Philosophy, Theosophy University Press, 2nd ed., 1979.

Prucker, G. de, Fountain-Source of Occultism, Theosophy University Press, 1974.

Prucker, G. de, Studies in Occult Philosophy, Theosophy University Press, 1945.

Judge, W.Q. Echoes of the Orient, Point Loma Publishing, 1975-1987.

Judge, W. Q., The Ocean of Theosophy, Theosophy University Press, 1973 (1893).

The Mahatma Letters to A. P. Sinnett, Theosophy University Press, 2nd ed., 1926.

(*) ***Available at TGS HiddenMysteries.com***

INTERNET REFERENCES TO THE INNER EARTH:

EXPLORATION AND IMAGINATION:
http://www.gei-net.com/users/v/vrartist/coconinocaves.html

MT SHASTA ANNOTATED BIBLIOGRAPHY:
http:www.siskiyou.edu/shasta/bib/B18.htm

AN AFTERNOON WITH MOTHER SEKHMET:
http://www.fourwinds 10.com/ari/transcriptions/sekhmet/sekhmet-091502.html

THE STARGATE CONSPIRACY:
http://www.cassiopaea.org/cass/stargate/htm

STARGATE CONSPIRACY
www.templarlodge.com

SUBTERRANEAN WORLDS INSIDE EARTH
www.comspiracyarchive.com/ufo/beckley/htm

THE ORGANIZATION; INSIDE THE COLLABORATION:
http://www.v-j-enterprises.com/undergrnd.html
http://www.v-j-enterprises.com/japicts.html

IT CAME FROM URANUS SAGAMEMMON'S TALE:
http://www.cinesight.com/itcame_from_urannus_Sag._Tale.htm

THE HOLLOW EARTH:
http://shamans-cave.com/maxpages/The_Hollow_Earth

TRAVELER TO THE INTERIOR OF THE HOLLOW EARTH:
http://onelight.com/colb/part1,2,3.html
onelight.com/hollow/books/earthscore.htm

SPIRIT WEB: THE HOLLOW EARTH:
http://www2.eu.spiritweb.org/Spirit/hollow-earth.html

DREAMLAND IN THE ROCKIES; THE SECRETS OF THE MOJAVE (Branton):
http://www.think-aboutit.com/branson/secrets of the mojave3.html

EARTH, MOTHER OF MYSTERIES; DEEP DWELLERS; RING AROUND THE OPENING:
Dean Dominic De Lucia
http://skywebsite.com/hollow/CCCC/id16.html
http://skywebsite.com/hollow/arctic/id15.html

MYSTERIES OF THE INNER EARTH (David Pratt):
THEOSOPHY AND THE INNER EARTH (Blavatsky):
http://ourworld.compuserve.com/homepages/dp5/inner4.html
http://ourworld.compuserve.com/homepages/dp5/hollow.html

PLATE TECTONICS: A PARADIGM UNDER THREAT:
http://ourworld.compuserve.com/homepages/dp5

ZEPHYR:
http://seekers.100megs6.com/Leslee%20Zephyr%20Astral.htm

OBSERVATIONS REPORTED WHILE ASTRAL TRAVELLING TO SKIROS IN THE AEGEAN SEA:
http://seekers.100megs6.com/Leslee%20skiros.htm

THEORIES OF THE HOLLOW EARTH:
http://www/ffiu.edu/mizrachs/hollow-earth.html

STRANGE AND UNEXPLAINED! - HOLLOW EARTH:
http://www.skygaze.com/content/strande/HollowEarth.shtlm

THE INCREDIBLE CITIES OF THE INNER EARTH (David H. Lewis):
http://www.hiddenmysteries.com/item4.html

HOLLOW EARTH:
http://yorkshireufoinfo.homestead.com/hollowearthg2.html

OPERATION POLARIS: International Society for a Complete Earth (I.S.C.E)
http://www.hollow-earth.org/polaris.html

THE ALLEGED DIARY OF ADMIRAL BYRD; OPERATION BYRD: (I.S.C.E.)
http://www.hollow-earth.org/diary.html
INNER EARTH BEINGS:
http://www.angelwater.net/y2k/inner_earth_beings.html

INNER EARTH:
http://www.angelfire.com/nc/HUMMINGBIRD1/inner/inner.html

UNDERGROUND TUNNEL ENTRANCES, INNER WORLDS, HOLLOW EARTH:
http://www.burlingtonnews.net/tunnels/html
http:www.sciencenews.org/sn_arch/7/20_96/fobl.html

THE HOLLOW EARTH, SUBTERRANEAN CIVILIZATION, AGHARTA:
http"//www.crystalinks.com/hollowearth.html

EARTH's GRIDS AND PORTALS:
http://www.mindbodyspirit.com.au/com/sol/soluntra8html

ANDEAN LEGENDS OF THE INNER EARTH:
http://www.labyrinthina.com/legend.html

SUBTERRANEAN TUNNELS AND THE HOLLOW EARTH:
http://sphinxtemple.virtualve.net/articles/tunnels2.html

THE SPECTRUM: RED ELK'S MESSAGE:
http://www.thespectrumnews.org/html/Red-Elk.html

PAO NEWSLETTER:
http://www.paoweb.som/newfo801.htm

MYSTERIES OF THE INNER EARTH:
http://www.nsf.gov/od/lpa/news/publict/frontier/3-97/3mysts.htm

A PROPHECY FROM THE INNER EARTH:
http://www.geocities.com/CapitolHill/6824/sacrgeo.htm

EXPLORATION AND IMAGINATION:
http://www.gci-net.com/users/v/vrartist/coconinocaves.html

THE ENTERPRISE MISSION (Michael Bara and Richard Hoagland):
"What is Happening at the South Pole?"
http://www.enterprisemission.com/millenn.htm

DULCE AND THE SECRET FILES OF A U. S. INTELLIGENCE WORKER:
http://www.title14.com/ufo/dulce/y5blupla.hjtml

THE ALIGNMENT GENERATION, review of John Major Jenkin's book
http://www.earthportals.com/Portal_Messenger/alignment2012

THE LEMURIAN INFLUENCE:
http://www.worldtrans.org/lyssa/tape130.html

THE LEMURIAN INFLUENCE, The Royal Priest Research:
http://www.worldtrans.org/lyssa/tape130.html

CITIES OF GOLD;
http://www.iamamerica.com/Pages/citiesofgold.html

GRAVITY OF INNER EARTH UPHEAVEALS:
http://www.huttoncommentaries.com/PSResearch/gravity_of_inner.htm

THE SMOKEY GOD: (Rodney M. Cluff)
ourholloweareth.com/SmokeyGod.htm

WORLD TOP SECRET: OUR EARTH IS HOLLOW, Rodney M. Cluff
http://www.ourhollowearth.com/HE/Earthbig.jpg
http://www.ourhollowearth.com/He/PolarOpn.htlm
http://www.ourhollowearth.com/HE/Uranus.html
http://www.ourhollowearth.com/HEGravity.html

healthresearchbooks.com/categories/hollowearth.htm

http://oneflyangel.com

Other books on this subject available at
TGS Publishers
hiddenmysteries.com

Hollow Planets by Jan Lamprecht

Hollow Earth Trilogy
The Hollow Earth
At the Earth's Core
The Smoky God

Caverns, Cauldrons and
Concealed Creatures
by Wm. Michael Mott

Incredible Cities of the Inner Earth
by David H. Lewis

The Cities of the Deep
by David H. Lewis

Hollow Earth in the Puranas
by Dean Dominic de Lucia

Etidorpha
John Uri Lloyd

Lost World of Agharti
by Alec MacLellan

Hollow Earth Enigma
by Alec MacLellan

Telos: The Call Goes Out from the
Hollow Earth
by Dianne Robbins

THE CALL GOES OUT From The Earth's Cetaceans
by Dianne Robbins

Messages From the Hollow Earth
by Dianne Robbins

Subterranean Worlds Inside Earth
by Timothy Green Beckley

Underground Alien Bases by Commander X

Zecharia Sitchin:

Are We Alone in the Universe?
The Cosmic Code
Divine Encounters
Genesis Revisited
The Lost Book of Enki
The Lost Realms
Of Heaven and Earth
Stairway to Heaven
12th Planet
Wars of Gods and Men
When Time Began

Vril
The Power of the Coming Race
by Edward George Bulwer-Lytton

Hidden Mysteries

TGS Publishers
HiddenMysteries.com
22241 Pinedale Lane
Frankston, Texas 75763

903-876-3256